MEMOIRS
of the
American Mathematical Society

ISSN 0065-9266 (print)
ISSN 1947-6221 (online)

Number 1321

AF621062

Existence of Unimodular Triangulations–Positive Results

Christian Haase
Andreas Paffenholz
Lindsay C. Piechnik
Francisco Santos

March 2021 • Volume 270 • Number 1321 (fifth of 7 numbers)

AMERICAN MATHEMATICAL SOCIETY

Library of Congress Cataloging-in-Publication Data

Cataloging-in-Publication Data has been applied for by the AMS.
See http://www.loc.gov/publish/cip/.
DOI: https://doi.org/10.1090/memo/1321

Memoirs of the American Mathematical Society

This journal is devoted entirely to research in pure and applied mathematics.

Subscription information. Beginning with the January 2010 issue, *Memoirs* is accessible from `www.ams.org/journals`. The 2021 subscription begins with volume 269 and consists of six mailings, each containing one or more numbers. Subscription prices for 2021 are as follows: for paper delivery, US$1085 list, US$868 institutional member; for electronic delivery, US$955 list, US$764 institutional member. Upon request, subscribers to paper delivery of this journal are also entitled to receive electronic delivery. If ordering the paper version, add US$20 for delivery within the United States; US$80 for outside the United States. Subscription renewals are subject to late fees. See `www.ams.org/help-faq` for more journal subscription information. Each number may be ordered separately; *please specify number* when ordering an individual number.

Back number information. For back issues see `www.ams.org/backvols`.

Subscriptions and orders should be addressed to the American Mathematical Society, P. O. Box 845904, Boston, MA 02284-5904 USA. *All orders must be accompanied by payment.* Other correspondence should be addressed to 201 Charles Street, Providence, RI 02904-2213 USA.

Copying and reprinting. Individual readers of this publication, and nonprofit libraries acting for them, are permitted to make fair use of the material, such as to copy select pages for use in teaching or research. Permission is granted to quote brief passages from this publication in reviews, provided the customary acknowledgment of the source is given.

Republication, systematic copying, or multiple reproduction of any material in this publication is permitted only under license from the American Mathematical Society. Requests for permission to reuse portions of AMS publication content are handled by the Copyright Clearance Center. For more information, please visit `www.ams.org/publications/pubpermissions`.

Send requests for translation rights and licensed reprints to `reprint-permission@ams.org`.

Excluded from these provisions is material for which the author holds copyright. In such cases, requests for permission to reuse or reprint material should be addressed directly to the author(s). Copyright ownership is indicated on the copyright page, or on the lower right-hand corner of the first page of each article within proceedings volumes.

Memoirs of the American Mathematical Society (ISSN 0065-9266 (print); 1947-6221 (online)) is published bimonthly (each volume consisting usually of more than one number) by the American Mathematical Society at 201 Charles Street, Providence, RI 02904-2213 USA. Periodicals postage paid at Providence, RI. Postmaster: Send address changes to Memoirs, American Mathematical Society, 201 Charles Street, Providence, RI 02904-2213 USA.

© 2021 by the American Mathematical Society. All rights reserved.

This publication is indexed in *Mathematical Reviews®*, *Zentralblatt MATH*, *Science Citation Index®*, *Science Citation Index™-Expanded*, *ISI Alerting Services℠*, *SciSearch®*, *Research Alert®*, *CompuMath Citation Index®*, *Current Contents®/Physical, Chemical & Earth Sciences.*
This publication is archived in *Portico* and *CLOCKSS*.
Printed in the United States of America.

∞ The paper used in this book is acid-free and falls within the guidelines established to ensure permanence and durability.
Visit the AMS home page at `https://www.ams.org/`

10 9 8 7 6 5 4 3 2 1 26 25 24 23 22 21

Contents

Abstract

Unimodular triangulations of lattice polytopes arise in algebraic geometry, commutative algebra, integer programming and, of course, combinatorics.

In this article, we review several classes of polytopes that do have unimodular triangulations and constructions that preserve their existence.

We include, in particular, the first effective proof of the classical result by Knudsen-Mumford-Waterman stating that every lattice polytope has a dilation that admits a unimodular triangulation. Our proof yields an explicit (although doubly exponential) bound for the dilation factor.

Received by the editor February 21, 2015, and, in revised form, December 21, 2017.

Article electronically published on June 24, 2021.

DOI: `https://doi.org/10.1090/memo/1321`

2010 *Mathematics Subject Classification.* Primary 52B20, 52B11; Secondary 13F20, 52C11.

The first author's affiliation is Freie Universität Berlin, Germany; haase@math.fu- berlin.de and was supported in part by NSF grant DMS-0200740, and by Emmy Noether HA 4383/1 and Heisenberg HA 4383/4 fellowships of the German Research Council (DFG).

The second author's affiliation is Technische Universität Darmstadt, Germany; email: paffenholz@mathematik.tu-darmstadt.de; and the second author was supported by a Priority Program (DFG/SPP 1489) of the German Research Council (DFG).

The third author's affiliation is Barnard College, NY; email: lpiechni@barnard.edu.

The fourth authors affiliation is the Universidad de Cantabria, Spain; email: francisco.santos@unican.es; and was supported in part by the Spanish Ministry of Science through grants MTM2011-22792 and MTM2014-54207-P, by the Alexander von Humboldt Foundation (grant 3.3-SPA/1147380 GSA), by the Einstein Foundation (grant EVF-2015-230), and by the National Science Foundation under Grant No. 1440140 while he was in residence at MSRI Berkeley in the fall of 2017.

©2021 American Mathematical Society

CHAPTER 1

Introduction

Unimodular triangulations of lattice polytopes arise in algebraic geometry, commutative algebra, integer programming and, of course, combinatorics. Admitting a unimodular triangulation is a property with nice implications, but presumably "most" lattice polytopes do not have such a triangulation (see Question 1.5 for a precise statement). In this paper we review methods for constructing unimodular triangulations and classes of polytopes which are known to have unimodular triangulations.

After defining the basic objects and fixing notation in Section 1.1, in Section 1.2 we explain why unimodular triangulations are important. We then embed the property of having a unimodular triangulation into a hierarchy of related properties in Section 1.2.5 and preview what does and does not appear in this paper in Sections 1.3 and 1.4, before listing some open questions in Section 1.5.

The rest of the paper is divided into three parts. In Section 2 we present methods of constructing unimodular triangulations, as well as how unimodular triangulations of certain polytopes can be used in constructions of unimodular triangulations for more complicated cases. This includes the analysis of joins, products, and projections. In Section 2.4 we review the relationship between unimodular triangulations and Gröbner bases of toric ideals, a correspondence that can be exploited in both directions.

Section 3 deals with particular classes of polytopes that can be shown to have unimodular triangulations and classes for which the question of existence has been addressed or is especially interesting. This includes polytopes related to root systems (Sections 3.1 and 3.2), polytopes related to graphs (Section 3.3) and smooth polytopes (Section 3.5). In particular, we show that all polytopes with facet normals in the root systems $\mathbf{B}_n$ have regular unimodular triangulations, a result that was not known before. (The same result for $\mathbf{A}_n$ easily follows from total unimodularity.) We also include previously unpublished results about empty smooth polytopes and smooth reflexive polytopes (e.g. Thm 3.32 and Thm 3.37).

Section 4 addresses dilations of lattice polytopes: when dilations must admit unimodular triangulations, and the process of constructing them. In particular, we provide the first effective version of the celebrated result of Knudsen, Mumford and Waterman saying that every lattice polytope P has a dilation cP that admits unimodular triangulations. Previous proofs of the theorem, although they can be considered algorithmic, do not mention an explicit bound for two reasons: (1) it is not easy to derive a bound from the algorithm, (2) the bound obtained would contain arbitrarily long towers of exponentials. Our bound, instead, is "only" doubly exponential:

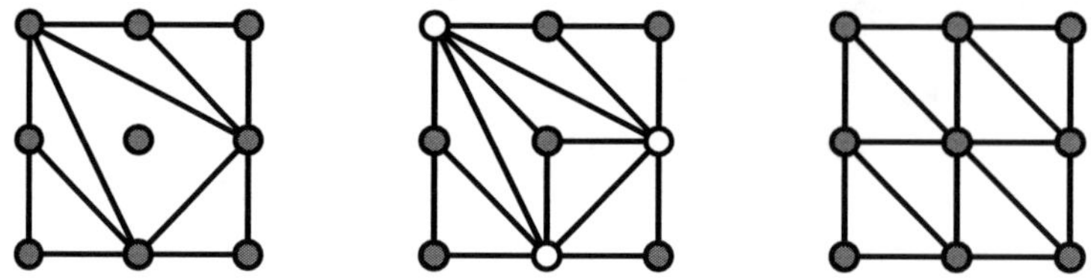

FIGURE 1.1. A non-unimodular triangulation and two unimodular ones. They are all regular, but only the last is quadratic. (See Section 2.4.2).

THEOREM 4.5. If P is a lattice polytope of dimension d and (lattice) volume $\mathrm{vol}(P)$, then the dilation

$$(d+1)!^{\mathrm{vol}(P)!(d+1)^{(d+1)^2 \mathrm{vol}(P)}} P$$

has a regular unimodular triangulation.

More precisely, if P has a triangulation $\mathcal{T}$ into N d-simplices, of volumes $V_1, \dots, V_N$, then the dilation

$$(d+1)!^{\sum_{i=1}^{N} V_i!\left((d+1)!^{d+1}\right)^{V_i-1}} \mathcal{T}$$

has a regular unimodular refinement.

1.1. What?

A *lattice polytope* in $\mathbb{R}^d$ is the convex hull of finitely many points in the lattice $\mathbb{Z}^d$. We identify two lattice polytopes if they are related by a lattice preserving affine map. Up to this *lattice equivalence*, we can always assume that our polytope is d-dimensional. (For more on convex polytopes and lattices we refer to [**9**].)

A *unimodular simplex* is a lattice polytope which is lattice equivalent to the standard simplex Δ^d, the convex hull of the origin $\mathbf{0}$ together with the standard unit vectors $\boldsymbol{e}_i$ $(1 \le i \le d)$. Equivalently, unimodular simplices are characterized as the d-dimensional lattice polytopes of minimal possible Euclidean volume, $1/d!$.

For the purposes of this paper, a (lattice) *subdivision* of a d-dimensional lattice polytope P is a finite collection of (lattice) polytopes $\mathcal{S}$ such that

(1) every face of a member of $\mathcal{S}$ is in $\mathcal{S}$,
(2) any two elements of $\mathcal{S}$ intersect in a common (possibly empty) face, and
(3) the union of the polytopes in $\mathcal{S}$ is P.

The maximal (d-dimensional) polytopes in $\mathcal{S}$ are called *cells* of $\mathcal{S}$.

A *triangulation* is a subdivision of a polytope for which each cell of the subdivision is a simplex. The triangulation is *unimodular* if every cell is. Figure 1.1 depicts three triangulations of the nine-point square. The first is not unimodular, while the other two are.

A *full triangulation* is a lattice triangulation which uses all the lattice points in P. The triangulation on the left in Figure 1.1 is not full. Every subdivision has a refinement to a full triangulation, for example the one resulting from the strong pulling procedure discussed in Section 2.1. Also, every unimodular triangulation is full, and in dimension at most two the converse is true as well. Depending on one's perspective, this is a consequence of, or the reason for, Pick's formula, which says that the area a polygon is one less than its number of interior lattice points

$$\mathsf{reeve}(q) \;:=\; \operatorname{conv}\begin{bmatrix} 0 & 0 & 0 & q \\ 0 & 0 & 1 & 1 \\ 0 & 1 & 0 & 1 \end{bmatrix}$$

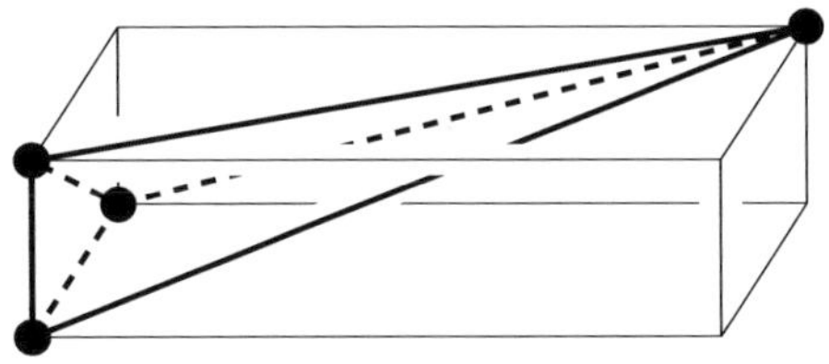

FIGURE 1.2. Reeve's tetrahedra $\mathsf{reeve}(q)$ for integral non-negative q.

plus half the number of lattice points on its boundary [**108**]. The formula yields the following proposition.

PROPOSITION 1.1. *Every lattice polygon has a unimodular triangulation.*

However, there are three dimensional polytopes for which a unimodular triangulation does not exist.

EXAMPLE 1.2 (John Reeve [**112**]). For $q \in \mathbb{Z}_{>0}$, the tetrahedron in Figure 1.2 contains only four lattice points — the vertices. Its only lattice triangulation is the trivial one. As the Euclidean volume is equal to $q/6$, this simplex does not have a unimodular triangulation for $q > 1$.

A subdivision is *regular* if its cells are the domains of linearity of a convex piecewise linear function. (Compare [**81**, Section 14.3], [**44**].) Less formally, a regular subdivision can be thought of as a subdivision that can be realized as a "convex folding" of the polytope (Figure 1.3 on the left). All three triangulations in Figure 1.1 are regular while the triangulation on the right in Figure 1.3 is not.

More formally, to define a regular subdivision of an arbitrary lattice polytope P (or to certify that a given subdivision is regular) we need a set of *weights* (or *heights*) $\boldsymbol{\omega} \in \mathbb{R}^{\mathcal{A}}$, where $\mathcal{A} = P \cap \mathbb{Z}^d$ is the set of lattice points in P. A lattice subpolytope F of P is a cell in the regular subdivision $\mathcal{S}_{\boldsymbol{\omega}}$ of P corresponding to those weights if there is a $\boldsymbol{\eta}_F \in \mathbb{R}^{\mathcal{A}}$ and a $\zeta_F \in \mathbb{R}$ such that

$$\omega_{\boldsymbol{a}} \geq \langle \boldsymbol{\eta}_F, \boldsymbol{a} \rangle + \zeta_F \quad \text{for all } \boldsymbol{a} \in \mathcal{A} \tag{1.1}$$

and

$$F = \operatorname{conv}\left\{\boldsymbol{a} \in \mathcal{A} \;:\; \omega_{\boldsymbol{a}} = \langle \boldsymbol{\eta}_F, \boldsymbol{a} \rangle + \zeta_F \right\}. \tag{1.2}$$

This can also be viewed geometrically. For this, consider the polyhedron $\tilde{P} = \operatorname{conv}(\boldsymbol{a} \times [\omega_{\boldsymbol{a}}, \infty) \;:\; \boldsymbol{a} \in \mathcal{A})$ in $\mathbb{R}^{d+1}$. The bounded faces of $\tilde{P}$ (which are also the lower faces if the last coordinate is considered as a height function) project to the faces of $\mathcal{S}_{\boldsymbol{\omega}}$. The latter are the domains of linearity of the function

$$\boldsymbol{x} \mapsto \min\bigl\{h \;:\; (\boldsymbol{x}, h) \in \tilde{P}\bigr\} = \max\bigl\{\langle \boldsymbol{\eta}_F, \boldsymbol{x} \rangle + \zeta_F \;:\; F \in \mathcal{S}_{\boldsymbol{\omega}}\bigr\}.$$

Conversely, we can of course prove that a given subdivision of a polytope is regular by finding an appropriate set of weights that generates the subdivision. A particular method for constructing regular full triangulations for an arbitrary lattice polytope is given in Lemma 2.1.

A *non-face* of a triangulation is a set of points whose convex hull does not form a face. Particularly important are the *minimal non-faces*, which do not form faces but for which every proper subset does. The list of minimal non-faces completely characterizes a triangulation, as does the list of (sets of points that form) cells. A

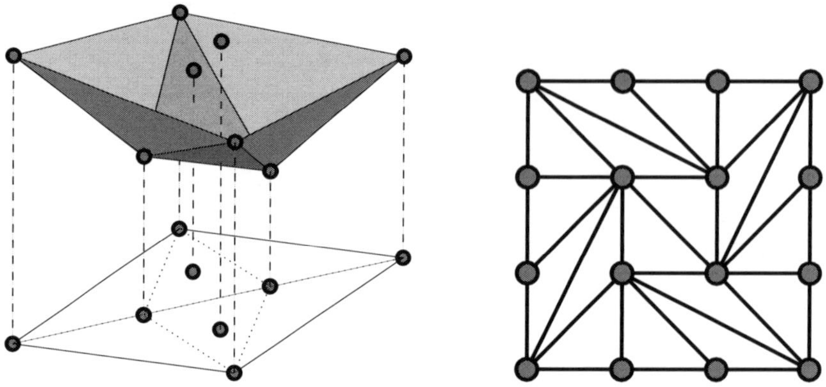

FIGURE 1.3. Regular vs. non-regular subdivisions.

triangulation for which all minimal non-faces contain only two elements is called *flag*. Putting these properties together, a *quadratic triangulation* is defined as a regular unimodular flag triangulation. The triangulation on the right in Figure 1.1 is quadratic. However, the three white vertices of the triangulation in the middle form a minimal non-face. So that triangulation is not quadratic.

1.2. Why? Who?

In this section, we present some applications of unimodular triangulations and closely related objects ("Why?"), arranged by mathematical discipline ("Who?").

1.2.1. Enumerative combinatorics. Many counting problems can be phrased as counting lattice points in (dilates) of polytopes or polyhedral complexes [**15, 41**]. By a fundamental result of Ehrhart, the number of lattice points in positive dilates kP of P is a polynomial function of degree d in $k \in \mathbb{Z}_{>0}$ [**15, 47**]. Consequently, the generating function has the special form:

$$\sum_{k\geq 0} \#(kP \cap \mathbb{Z}^d)\, t^k = \frac{h^*(t)}{(1-t)^{d+1}},$$

where $h^*(t)$ is a polynomial of degree $\leq d$. If P has a unimodular triangulation, h^* equals the combinatorial h-polynomial of that triangulation [**19**]. This means if different unimodular triangulations exist they have the same f-vector.

For example, the chromatic polynomial of a graph can be interpreted as the Ehrhart polynomial of the complement of a hyperplane arrangement in the 0/1-cube [**16**]. This hyperplane arrangement is compatible with the standard unimodular triangulation of the cube as the order polytope of an anti-chain (see Section 3.1.1). This can be used to compute the chromatic polynomial in terms of Steingrímsson's coloring complex [**128**]. Hersh and Swartz used this to derive rather strong conditions for a polynomial to be a chromatic polynomial of a graph [**63**]. A very similar argument applies for polynomials counting nowhere-zero flows or nowhere-zero tensions [**23**] and even for the Tutte polynomial of a graph [**24**].

In combinatorial representation theory, Littlewood-Richardson coefficients in type A, and more general Clebsch Gordan coefficients in other types, can be represented as the number of lattice points in specific polytopes ([**17, 18**] and [**34, 78**]).

In type A, Knutson and Tao prove the Saturation Theorem that implies Horn's Conjecture by showing that the hives polytope has an integral vertex whenever it is non-empty. This result would also follow from De Loera and McAllister's conjecture [**43**, Conj. 4.5] that the so called homogenized hive matrix has a unimodular triangulation, which they have validated up to A_6.

Other well known enumerative results involving unimodular triangulations include the proof of the hook-length formula by Pak [**105**] and Stanley's observation that Eulerian numbers are volumes of hypersimplices [**52, 125**] (see also Section 3.1.1).

1.2.2. Integer programming. Embedding a polytope at height one as $P \times \{1\} \subset \mathbb{R}^{d+1}$ generates a pointed polyhedral cone $\sigma_P \subset \mathbb{R}^{d+1}$. The *Hilbert basis*, $\mathcal{H}$, of this cone is the minimal generating set for the semigroup $\sigma_P \cap \mathbb{Z}^{d+1}$. $\mathcal{H}$ clearly contains the set of lattice points in $P \times \{1\}$. If the converse holds then P is called *integrally closed* or to have the *integer decomposition property*. A sufficient condition for this to happen is that P has a unimodular triangulation or, more generally, that P can be covered by unimodular simplices. In fact, there is a hierarchy of covering properties interpolating between integrally closed polytopes and polytopes with unimodular triangulations (see Section 1.2.5).

This Hilbert basis hierarchy appears in integer programming in two guises: test sets and TDI systems. For the first, suppose we want to solve, for a fixed matrix $A \in \mathbb{Z}^{d \times n}$, the family of integer programs $\mathrm{IP}(\boldsymbol{b}; \boldsymbol{c}; \boldsymbol{u})$ given by

$$\min \{ \langle \boldsymbol{c}, \boldsymbol{x} \rangle \ : \ \boldsymbol{x} \in \mathbb{Z}^n, \ A\boldsymbol{x} = \boldsymbol{b}, \ \boldsymbol{0} \le \boldsymbol{x} \le \boldsymbol{u} \}$$

for varying $\boldsymbol{b} \in \mathbb{Z}^d$, $\boldsymbol{u} \in \mathbb{Z}^n$, $\boldsymbol{c} \in (\mathbb{Z}^n)^*$. For every sign pattern $\boldsymbol{\varepsilon} \in \{\pm 1\}^n$ consider the pointed polyhedral cone

$$\sigma_{\boldsymbol{\varepsilon}} = \{ \boldsymbol{x} \in \mathbb{R}^n \ : \ A\boldsymbol{x} = \boldsymbol{0}, \ \varepsilon_i x_i \ge \boldsymbol{0} \ \text{ for } \ i = 1, \dots, n \}$$

together with its Hilbert basis $\mathcal{H}_{\boldsymbol{\varepsilon}}$. Then the *Graver basis* $\bigcup \mathcal{H}_{\boldsymbol{\varepsilon}}$ is a test set for our family of IPs in the sense that for every feasible non-optimal point we find an improving vector in our finite test set [**2, 42, 62, 122**].

TDI systems [**120**, §22] are closely related. A system of linear inequalities $A\boldsymbol{x} \le \boldsymbol{b}$ with integer coefficients is called *totally dual integral* (*TDI*) if for every $\boldsymbol{c}$ such that the dual linear program (LP)

$$\min\{ \langle \boldsymbol{y}, \boldsymbol{b} \rangle \ : \ \boldsymbol{y}A = \boldsymbol{c}, \ \boldsymbol{y} \ge \boldsymbol{0} \}$$

is bounded, there exists an integral dual optimal solution.

This property is equivalent to the condition that the active constraints for every face of the feasible polyhedron

$$P \ = \ \{ \boldsymbol{x} \ : \ A\boldsymbol{x} \le \boldsymbol{b} \}$$

form a Hilbert basis of the normal cone [**120**, §22.3]. This is guaranteed, for example, if the normal fan of P can be refined to a regular triangulation using no additional rays than the rows of A. If $A\boldsymbol{x} \le \boldsymbol{b}$ is TDI and $\boldsymbol{b}$ is integral then P is a lattice polytope.

A particularly nice special class of TDI systems are those with a totally unimodular constraint matrix A. A matrix $A \in \mathbb{R}^{m \times n}$ is *unimodular* if every maximal minor is in $\{0, \pm 1\}$, and it is *totally unimodular* if this holds for all minors. Note that negating columns or rows and adding unit vectors to the matrix does not change total unimodularity. There are various criteria to check whether a given matrix A is totally unimodular, see e.g. [**120**, §19.2] for a list and references. A system $A\boldsymbol{x} \leq \boldsymbol{b}$ is TDI for every integral right hand side $\boldsymbol{b}$ if and only if A is totally unimodular.

Surprisingly, many interesting families of polytopes can be defined by totally unimodular constraint matrices. In fact, incidence matrices of bipartite graphs, incidence matrices of directed graphs, and sub-configurations of the root system $\mathbf{A}_{n-1}$ are totally unimodular. For the former see Section 3.3.2, and for the latter class, which includes order polytopes and hypersimplices see Section 3.1.1. Furthermore, we consider polytopes spanned by the columns of incidence matrices in Sections 3.2.1 and 3.2.3.

Suppose we want to minimize a linear functional $\boldsymbol{c} \in \mathbb{R}^n$ subject to constraints $A\boldsymbol{x} = \boldsymbol{b}$, and $\boldsymbol{x} \geq \mathbf{0}$. We can interpret the columns of the $d \times n$ matrix A as vectors in $\mathbb{R}^d$. Clearly, the system is (LP-)feasible if and only if $\boldsymbol{b}$ belongs to the cone generated by these columns of A. If we assume that this cone is pointed, then we can consider c_i as a weight of the i-th vector so that $\boldsymbol{c}$ induces a regular subdivision of the cone. If this subdivision happens to be a unimodular triangulation, then the LP-optimum equals the IP-optimum for any $\boldsymbol{b}$. This is because a feasible $\boldsymbol{b}$ has a non-negative integral representation in "its" cone, and the PL function induced by $\boldsymbol{c}$ is convex.

1.2.3. Commutative algebra. Many properties of combinatorial objects have direct translations to algebraic objects like semigroup algebras, monomial ideals, toric varieties, and singularities, via the correspondence

lattice point		Laurent monomial
$\boldsymbol{a} = (a_1, \ldots, a_d) \in \mathbb{Z}^d$	$\longleftrightarrow$	$\boldsymbol{x}^{\boldsymbol{a}} := x_1^{a_1} \cdot \ldots \cdot x_d^{a_d} \in \mathbb{K}[x_1^{\pm 1}, \ldots, x_d^{\pm 1}]$

Commutative algebraists are interested in the properties of the graded semigroup ring $R_P = \Bbbk[\sigma_P \cap \mathbb{Z}^{d+1}]$. For example, P is integrally closed if and only if the domain R_P is generated in degree one.

Normality is a closely related notion. Consider the subring $\tilde{R}_P \subseteq R_P$ generated by the degree one piece. P is called *normal* if $\tilde{R}_P$ is normal (integrally closed in its quotient field). That is, P is normal if $\Bbbk[\sigma_P \cap \Lambda]$ is generated in degree one, where $\Lambda \subseteq \mathbb{Z}^{d+1}$ is the sublattice generated by $(P \times \{1\}) \cap \mathbb{Z}^{d+1}$ [**27**, Def. 2.59].

Further, there is a close connection between the Gröbner bases of the defining ideal $\mathcal{I}_P$ of $\tilde{R}_P = \Bbbk[x_1 \ldots x_n]/\mathcal{I}_P$ ($n = |P \cap \mathbb{Z}^d|$) and regular triangulations of P. In particular, if P has a regular unimodular triangulation $\mathcal{T}$, then $R_P = \tilde{R}_P$ and, moreover, there is a Gröbner consisting of binomials and with leading terms given by the minimal non-faces of $\mathcal{T}$ (this correspondence is demonstrated in Section 2.4). The degree of each binomial in the Gröbner basis is equal to the size of the corresponding non-face. So, triangulations provide degree bounds for Gröbner bases. This makes the search for quadratic triangulations of particular interest, as the existence of such a triangulation guarantees the existence of a quadratic Gröbner basis which in turn shows that the algebra $\tilde{R}_P = R_P$ is Koszul (i.e. $\Bbbk$ has a linear free resolution as an R_P-module) [**30, 48**].

1.2.4. Algebraic geometry. Algebraic geometry associates two objects with P [**50, 53, 93**]. First there is the affine Gorenstein toric variety

$$U_P = \operatorname{Spec} \Bbbk[\sigma_P^\vee \cap \mathbb{Z}^{d+1}],$$

where $\sigma_P^\vee$ is the cone dual to σ_P. Here a unimodular triangulation of P corresponds to a crepant desingularization of U_P, which is projective if and only if the triangulation is regular. These crepant birational morphisms have been used to reduce canonical singularities to $\mathbb{Q}$-factorial terminal singularities, and to treat minimal models in high dimensions. They appear in the high-dimensional McKay correspondence [**67**, **113**] for Gorenstein quotient singularities $\mathbb{C}^d/G$, proven by Batyrev [**10**]. Moreover, a one-to-one correspondence of McKay-type also holds for triangulation induced resolutions of U_P [**11**].

One of the earliest, and to this day one of the most striking, results involving unimodular triangulations is the stable reduction theorem of Kempf, Knudsen, Mumford and Saint-Donat [**75**]. They showed that in characteristic zero, every one-dimensional family can be resolved so that the exceptional locus is a normal crossing divisor. They achieved this by reducing the statement to the case of "toroidal" singularities. The combinatorial core of the argument is the statement that every lattice polytope has some dilation admitting a regular unimodular triangulation [**77**]. This is the content of Theorem 4.5, which we discuss in detail in Section 4.

The other algebraic geometric object associated with P is the projectively embedded toric variety

$$X_P = \operatorname{Proj} \mathbb{k}[\sigma_P \cap \mathbb{Z}^{d+1}] = \operatorname{Proj} R_P.$$

Here the polytope P specifies an ample line bundle L_P on X_P. (see [**53**, Section 3.4].) If X_P is smooth (i.e., the normal fan of P is unimodular), then L_P is very ample, and provides an embedding $X_P \hookrightarrow \mathbb{P}^{n-1}$, where $n = \#(P \cap \mathbb{Z}^d)$. The two following questions about the defining equations of a smooth $X_P \subset \mathbb{P}^{n-1}$ have been around for quite a while, but the origins are hard to track.

First, if P is a lattice polytope whose corresponding projective toric variety is smooth, is the defining ideal $\mathcal{I}_P$ generated by quadrics? And secondly, is the embedding $X_P \hookrightarrow \mathbb{P}^{n-1}$ of a smooth X_P *projectively normal* (i.e. is R_P generated in degree one)?

Both questions have a positive answer for polytopes with a quadratic triangulation (see Section 2.4.2), both questions have a negative answer without the smoothness assumption. (See Knutson's section in [**14**] for a historical discussion and partial results.)

1.2.5. A hierarchy of covering properties. The property of admitting a unimodular triangulation embeds into a large hierarchy of algebraic and convex geometric properties. We list some of the more combinatorial ones in decreasing strength. Compare [**88**, p. 2097f] and [**89**, p. 2313ff].

(1) $P \cap \mathbb{Z}^d$ is totally unimodular
(2) P is compressed (see Section 2.1)
(3) P has a regular unimodular triangulation
(4) P has a unimodular triangulation
(5) P has a unimodular binary cover (a cycle generating $H_d(P, \partial P; \mathbb{Z}_2)$ formed by unimodular simplices)
(6) P has a unimodular cover
(7) σ_P is integrally closed and has a free Hilbert cover (every lattice point is a $\mathbb{Z}_{\geq 0}$-linear combination of linearly independent lattice points in $P \times \{1\}$)

(8) P is integrally closed

All the implications $(i) \Rightarrow (i+1)$ are strict [**25**]. See [**51**] for other covering properties.

Property (7) is known to be equivalent to σ_P being integrally closed and satisfying the *integral Caratheodory property* [**25**]: every lattice point is a $\mathbb{Z}_{\geq 0}$-linear combination of $\dim C$ many lattice points in $P \times \{1\}$).

Most of this hierarchy's properties have direct translations into an algebraic language about R_P and $\mathfrak{I}_P$. However, there are a few additional algebraic properties which do not quite fit and have their own shorter hierarchy.

(1') P has a quadratic triangulation (see §2.4.2)
(2') $\mathfrak{I}_P$ has a quadratic Gröbner basis
(3') R_P is a Koszul algebra ($\Bbbk$ has a linear free resolution as an R_P-module)
(4') $\mathfrak{I}_P$ is generated by quadrics

The two hierarchies are linked by the fact that a quadratic triangulation is a regular unimodular triangulation.

1.3. What is new?

In this section we will give a brief summary of the results contained in this paper that have not appeared elsewhere, or not appeared in this form.

1.3.1. Section 2. In Section 2.1 we clarify the relationship between two notions of pulling that have been used in the literature, and sometimes confused with one another. We also show a general procedure for obtaining regular, unimodular and/or flag triangulations of (some) lattice polytopes, by first dicing by hyperplanes and then refining. We show that all polytopes with unimodular facet matrices have regular, unimodular triangulations. However, the (3×3)-Birkhoff polytope shows that not all have quadratic ones.

Section 2.2 introduces pull-back and push-forward techniques for constructing unimodular triangulations. These methods were announced in [**60**]. Here we give extended versions of the constructions with full proofs. We use this to construct regular unimodular triangulations for smooth reflexive polytopes. An update on these results is given in Section 3.5.2.

We prove that (regular and unimodular) triangulations of polytopes can be extended to give (regular and unimodular) triangulations of their products, joins, and some other constructions (Section 2.3). In particular, we rework and extend to the case of non regular triangulations a result of Ohsugi and Hibi about nested configurations, for which we introduce the notion of a semidirect product of polytopes [**103**]. We also extend a result of Sullivant [**135**] on toric fiber products to the case of positive codimension.

Section 2.4 contains a direct proof of the correspondence between regular unimodular triangulations and square-free initial ideals. Finally, in Section 3.6 we briefly recall the relation between toric Hilbert schemes, Gröbner bases and unimodular triangulations. We briefly reproduce the example of a disconnected toric Hilbert scheme by Santos, based in the root configuration of type $\mathbf{F}_4$.

1.3.2. Section 3. Here we give several new examples of families of polytopes that do (or do not) have regular unimodular triangulations, and some with additional properties.

We start by considering polytopes whose facet normals belong to a root system (Section 3.1). Those of type **A** can be triangulated via hyperplane dicing, which implies that they have quadratic triangulations. For type **B** we show that the hyperplane subdivision by short roots can be extended to produce a regular unimodular triangulation (but not necessarily a quadratic one). In contrast, we show examples of polytopes cut out by hyperplanes of type $\mathbf{E}_8$ and $\mathbf{F}_4$ and which do not have unimodular triangulations. In fact, these examples are not even integrally closed.

Section 3.2 compiles several results about polytopes whose vertices belong to a root system, giving a unified approach to and sometimes shorter proofs of known results.

Smooth (3×3)-transportation polytopes have been studied in [**61**] using hyperplane subdivision. Here we extend this with a theorem of Lenz proving that all simple 3×4 transportation polytopes have quadratic triangulations (Proposition 3.26). In Section 3.5.1 we show that smooth empty lattice polytopes are products of unimodular simplices.

1.3.3. Section 4. This section revolves around dilations of lattice polytopes, including the proof of our effective version of the Knudsen-Mumford-Waterman Theorem (Theorem 4.5). As the original one, our proof is based on induction on the maximum volume of a simplex in an initial triangulation of P

Before that, in Section 4.2 we study refinements of dilations of unimodular triangulations. In particular, we show a new procedure to quadratically refine a dilation of a quadratic triangulation (Theorem 4.8).

1.4. What is not here

Although we try to give a thorough overview of methods for obtaining unimodular triangulations, we cannot cover everything in this paper. In particular, we restrict our attention to (geometric) triangulations of lattice polytopes. We do not cover subdivisions of cones or combinatorial abstractions via oriented matroids. We also do not consider methods for modifying a triangulation via flips.

There are a number of computational tools available to test for and explore triangulations, secondary fans, etc. Among the prominent tools are the software package `normaliz2` of Bruns, Ichim, and Söger [**31, 33**] for the computation of normalizations of cones; `4ti2` of Hemmecke, Köppe, and DeLoera [**1**] for the computation of generating sets of integer points in polyhedra, Hilbert and Gröbner bases; `TOPCOM` by Rambau [**111**] for various types of triangulations; `LattE Integrale` by De Loera et al. for computing Ehrhart series; The software `polymake` [**54, 109**] offers fast enumeration of lattice points and extensive computations with polyhedra, fans, and triangulations. For the toric methods explained in Section 2.4 one should also consider the computer algebra packages `Singular` [**45**], `Macaulay2` [**57**], and CoCoA [**39**]. The software `polymake` offers many computations for toric ideals and varieties via extensions [**73, 74**].

Finally, this survey also does not comment on connections between the emerging field of tropical geometry and lattice polytopes, their subdivisions, and the secondary fan. For this, see for example [**68**].

1.5. What is left?

To close this introduction we here compile several open questions on (regular, unimodular, or flag) triangulations of lattice polytopes. One general open question is: how special is it for a lattice polytope to have a unimodular triangulation? One way to make this precise is as follows.

QUESTION. *For a given fixed dimension d and volume V consider the set of lattice d-polytopes of normalized volume at most V, modulo unimodular equivalence, which is a finite set* [**5, 7, 8**].

Let $u(d, V) \in [0, 1]$ denote the fraction of them that admit unimodular triangulations. We conjecture that

$$\limsup_{V\to\infty} \sqrt[V]{u(d,V)} < 1,$$

for every fixed $d \geq 3$.

Other questions we think are important are the following.

1.5.1. Smoothness and normality, vs. unimodular triangulations. The following two questions have been mentioned already in Section 1.2.4.

- Does every smooth polytope have a unimodular triangulation, or at least a unimodular cover? A positive answer to this would imply a positive answer to the following very prominent open problem by Oda: Is every smooth projective toric variety *projectively normal*? Both questions are open even in dimension three. Various partial results have been proved in this direction (see [**58, 61, 82, 89**], and computational approaches in [**59, 84**]).
- Is the ideal of every smooth projective toric variety generated by quadrics? Here, a positive answer would follow from the existence of a quadratic triangulation for every smooth polytope.

More generally, for every property of the hierarchies in Section 1.2.5 above one can ask whether or not all smooth polytopes satisfy that property. Every one of these questions is open in dimensions three and above. In Section 3.5.2 we investigate smooth reflexive polytopes which have been classified up to dimension nine. Of the 80 892 smooth reflexive polytopes in dimension at most seven, there are 18 for which we could not decide the existence of a quadratic triangulation – two in dimension six and 16 in dimension seven.

1.5.2. Polytopes cut out by roots. Athough we here considerably extend the results on triangulations of polytopes cut out by root systems (see Section 3.1), some cases are still open. In particular, it is not known whether polytopes whose facet normals are contained in $\mathbf{C}_n$ or $\mathbf{D}_n$ have unimodular triangulations (we expect that they do), and the same for those with facet normals in $\mathbf{E}_7$ or $\mathbf{E}_6$ do (we expect that they do not; in fact we do not expect them to be integrally closed, in general). Further, although we show that $\mathbf{B}_n$-polytopes have a regular and unimodular triangulation, we do not know whether they have quadratic ones.

Another natural question is as follows: the way we define polytopes cut out by roots is as those whose normal vectors are in the root system. But one can also restrict further and look at polytopes whose normal fan is refined by the cluster complex for that root system. If the normal fan agrees with the cluster complex, this is a particular question of the smoothness question above, since generalized associahedra have simplicial and unimodular normal fans. But it is also a question that has more chances to be answered in the positive, since there is the additional machinery of root systems and cluster algebras available. In type **A**, this class contains the class of matroid polytopes and thus could shed some light on an old conjecture of Neil White [**138**] that the toric ideal of a matroid polytope is generated in degree two. A partial positive answer is in [**79**], but a quadratic triangulation of the matroid polytope would even yield a quadratic Gröbner basis.

1.5.3. Dilated polytopes. There are several open questions concerning the dilation factors c that make the dilation cP have a unimodular triangulation. Among them we can mention the following:

(1) Is it true that if c_1P and c_2P have unimodular triangulations, then $(c_1 + c_2)P$ has one? That is, is the set of valid dilation factors a subsemigroup of $\mathbb{N}$? This is known to hold for empty simplices in dimension three [**116**], but not in general.
(2) Is it true that for every P there is a c_0 such that cP has a unimodular triangulation for all $c \geq c_0$? This is Problem 5 in [**29**]. Remember that it is not always true that if cP has a (regular, flag) unimodular triangulation then $(c+1)P$ has one too (see Example 4.13)
(3) The bound in Theorem 4.5 is doubly exponential both in the dimension and the volume of the starting polytope. Is there a singly exponential one? Is there a bound depending on dimension but not on volume (a bound "constant in fixed dimension")? Is the latter true at least in dimension four?

    We do not believe a global bound (in fixed dimension) to exists for $d \geq 5$. However, given a triangulation of a d-polytope P in which no simplex has volume greater than V, we believe that a dilation factor c of about d^V should be sufficient for cP to have a unimodular triangulation.

All these questions have a positive answer if we only ask for unimodular *covers* rather than triangulations. In this case for every lattice polytope P there is a threshold c_P such that cP has a unimodular cover if and only if $c \geq c_P$. Different polytopes may have different thresholds, but for every dimension there is a common c_d upper bounding the thresholds of all d-polytopes and depending polynomially on d [**27**, Theorem 3.23].

We remark that for the IDP property there are polytopes where the valid dilations do not form a semigroup: it follows from [**80**] that for each k there is a $(2k-1)$-dimensional lattice polytope P such that cP is IDP if, and only if, c is not a proper divisor of k. For example, there is a 49-dimensional polytope such that $2P$ and $3P$ are IDP but $5P$ is not. See details in Example 4.14.

1.5.4. Other questions. Lattice parallelepipeds and, more generally, lattice zonotopes are known to be IDP. Is it true that they all have unimodular triangulations?

Lecture hall simplices (see Section 3.4) arise very naturally in combinatorics and have relations to partitions of integers. In their original form, LHS_{d+1} has recently been shown to have a quadratic triangulation. But for their generalizations $\mathrm{LHS}_{d+1}(\boldsymbol{s})$ (see (3.2)) it is not known what sequences $\boldsymbol{s} = (s_i)_{i=1}^d$ yield polytopes admitting unimodular triangulations. More generally, we ask which $\boldsymbol{s}$-order polytopes $O(\preccurlyeq, \boldsymbol{s})$ have unimodular triangulations.

Do the homogenized hive matrices of De Loera and McAllister [**43**] always have unimodular triangulations? (see Section 1.2.1 for details).

CHAPTER 2

Methods

2.1. Pulling Triangulations

Pulling refinements are a useful tool for constructing regular triangulations. Two distinct versions of pulling refinements appear in the literature. We will call them weak pulling and strong pulling.

2.1.1. Weak and Strong Pulling. We first discuss strong pulling. If $\mathcal{S}$ is a subdivision of P and $\boldsymbol{m} \in P \cap \mathbb{Z}^d$ is a lattice point in P, the *strong pulling refinement* $\operatorname{pull}_{\boldsymbol{m}} \mathcal{S}$ is obtained by replacing every face $F \in \mathcal{S}$ containing $\boldsymbol{m}$ by the pyramids $\operatorname{conv}(\boldsymbol{m}, F')$, for each face F' of F which does not contain $\boldsymbol{m}$.

Here are some facts about the structure of pulling subdivisions.

LEMMA 2.1. *Let P be a lattice polytope with a subdivision $\mathcal{S}$.*

(1) *Strong pulling preserves regularity.*
(2) *Strongly pulling all lattice points of P in any order results in a full triangulation.*
(3) *If only vertices of P are pulled, then every maximal cell of the refinement will be the join of the first pulled vertex $\boldsymbol{v}_1$ with a maximal cell in the pulling subdivisions of the facets not containing $\boldsymbol{v}_1$.*

In particular, this Lemma guarantees that every (regular) lattice subdivision of a lattice polytope has a (regular) refinement which is a full triangulation.

PROOF. *(1)*: Let $\mathcal{S}$ be the regular subdivision of P given by a weight vector $\boldsymbol{\omega} \in \mathbb{R}^{\mathcal{A}}$ ($\mathcal{A} = P \cap \mathbb{Z}^d$). By definition of regularity, the lower faces of the lifted polyhedron $\tilde{P} = \operatorname{conv}(\boldsymbol{a} \times [\omega_{\boldsymbol{a}}, \infty) \ : \ \boldsymbol{a} \in \mathcal{A})$ in $\mathbb{R}^{d+1}$ project to $\mathcal{S}$. Let $\boldsymbol{m} \in \mathcal{A}$ and set $\omega'_{\boldsymbol{m}} = \min\{h \ : \ (\boldsymbol{m}, h) \in \tilde{P}\} - \epsilon$ and $\omega'_{\boldsymbol{a}} = \omega_{\boldsymbol{a}}$ for all $\boldsymbol{a} \in \mathcal{A} \setminus \{\boldsymbol{m}\}$. Then, for small enough $\epsilon > 0$, the pulling refinement $\operatorname{pull}_{\boldsymbol{m}}(\mathcal{S})$ is induced by the weights $\boldsymbol{\omega}'$.

(2): Every face of $\operatorname{pull}_{\boldsymbol{m}}(\mathcal{S})$ containing $\boldsymbol{m}$ is a pyramid with apex $\boldsymbol{m}$. If $Q \in \mathcal{S}$ has $\boldsymbol{n}$ as an apex, then every face of $\operatorname{pull}_{\boldsymbol{m}}(\mathcal{S})$ inside Q and containing $\boldsymbol{n}$ still has $\boldsymbol{n}$ as an apex. After strongly pulling all lattice points, each lattice point is a vertex of the subdivision, and every vertex of every cellos an apex. So, each cell is a simplex.

(3): If we apply the previous argument to the trivial subdivision of P, $\boldsymbol{v}_1$ must be an apex of every cell. □

The other notion of pulling, weak pulling, arises in the context of subdivisions of point configurations [**44**] (or, equivalently, subdivisions of marked polytopes [**56**]). Basically, in a subdivision of a point configuration each face $F \in \mathcal{S}$ is determined by the convex hull of some (possibly full) subset of its lattice points, and weak pulling treats lattice points in this set differently than those not in this set for each F.

For example, consider regular subdivisions $\mathcal{S}_1$ and $\mathcal{S}_2$ of the one-dimensional configuration $\mathcal{A} = \{1, 2, 3, 4\} \subset \mathbb{Z}^1$ obtained from the weight vectors $\boldsymbol{\omega}_1 = (0, 0, 0, 1)$

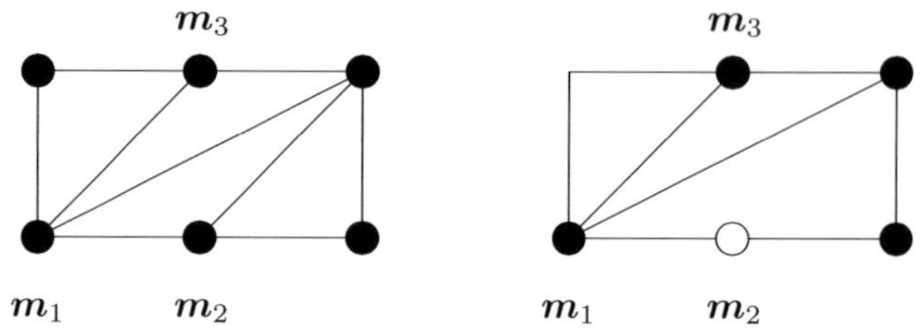

FIGURE 2.1. Strongly and weakly pulling $\boldsymbol{m}_1$, then $\boldsymbol{m}_2$, then $\boldsymbol{m}_3$.

and $\boldsymbol{\omega}_2 = (0, 1, 0, 1)$ respectively. They both consist of the segments $[1, 3]$ and $[3, 4]$ but in $\mathcal{S}_1$ the segment $[1, 3]$ appears as the convex hull of 1, 2 and 3 while in $\mathcal{S}_2$ it appears as the convex hull of 1 and 3 alone. When looking at the set of all subdivisions of $\mathcal{A}$ as a whole, there are good reasons to consider $\mathcal{S}_1$ and $\mathcal{S}_2$ as different, $\mathcal{S}_1$ not being a triangulation and $\mathcal{S}_2$ being a triangulation that refines $\mathcal{S}_1$.

In this context, when performing a pulling at $\boldsymbol{m}$ it is natural to remove the lattice points in each pyramid $\operatorname{conv}(\boldsymbol{m}, F')$ from the list of points that can be pulled later, but not those in its base F'. This is what we call *weak pulling*. In the setting of toric algebra as seen in Section 2.4, the weak pulling triangulation corresponds to the reverse lexicographic term order. Figure 2.1 illustrates the difference of the two versions.

The following is the analogue of Lemma 2.1 for weak pullings. For a proof we refer to [**44**, Section 4.3.4]:

LEMMA 2.2. *Let P be a lattice polytope with a subdivision $\mathcal{S}$.*
(1) *Weak pulling preserves regularity.*
(2) *Weakly pulling all lattice points of P in any order yields a triangulation.*

In this article, we will only use pulling refinements in cases where all lattice points in the polytope P are vertices of the subdivision $\mathcal{S}$. With this assumption, strong and weak pulling subdivisions agree, so there is no ambiguity when referring to *pulling subdivisions.*

2.1.2. Compressed Polytopes. Stanley calls a polytope *compressed* if all its weak pulling triangulations are unimodular [**126**]. This clearly implies that the only lattice points in P are its vertices. Because of Theorem 2.3 below, compressed polytopes are sometimes called *2-level* polytopes.

Surprisingly many well-known polytopes fall into this category. Examples of compressed polytopes include the Birkhoff polytope (Section 3.3.1), order polytopes and hypersimplices (Section 3.1.1), stable set polytopes of perfect graphs (Section 3.3.2), and the integer hulls of weighted Gelfand-Tsetlin polytopes $P_{\lambda/\mu,\mathbf{1}}$ for weight $\mathbf{1}$ [**3**]. Athanasiadis was even able to use the fact that the Birkhoff polytope is compressed to prove unimodality of its h^*-vector [**6**] (see also [**32**]).

There are several characterizations of compressed polytopes. Here we present one based on width with respect to facets. If P is a lattice polytope and $\langle \boldsymbol{y}_i, \boldsymbol{x} \rangle \geq c_i$ for $i = 1, \ldots, n$ are the facet defining inequalities with primitive integral $\boldsymbol{y}_i$, the *width* of P with respect to the i-th facet (or with respect to the direction of $\boldsymbol{y}_i$) is the difference

$$\max\langle \boldsymbol{y}_i, P \rangle \; - \; \min\langle \boldsymbol{y}_i, P \rangle.$$

In particular, P has width one with respect to a facet if it lies between the hyperplane spanned by this facet and the next parallel lattice hyperplane.

The main implication of the following theorem is due to the fourth author. The proof we present here is the original one (MSRI 1997, unpublished). It was subsequently also proven by Ohsugi and Hibi [**99**] and by Sullivant [**134**].

THEOREM 2.3. *Let P be a lattice polytope. Then the following are equivalent:*

(1) *P is compressed.*
(2) *P has width one with respect to all its facets.*
(3) *P is lattice equivalent to the intersection of a unit cube with an affine space.*

PROOF. (2) $\Longrightarrow$ (1): By decreasing induction on the dimension one sees that every face of P has width one with respect to all facets. The restriction of a pulling triangulation to any face is a pulling triangulation itself and thus unimodular (by another induction). Hence, every maximal simplex in the triangulation of P is the join of a unimodular simplex in some facet with the first lattice point that was pulled.

The other implications are easy. □

2.1.3. Hyperplane Arrangements. In this section, we apply the above characterization of compressed polytopes to triangulate "bigger" polytopes using hyperplane arrangements.

Let $\mathcal{A} := \{\boldsymbol{n}_1, \ldots, \boldsymbol{n}_r\} \subset \mathbb{Z}^d$ be a collection of vectors that span $\mathbb{R}^d$ and form a unimodular matrix, (i.e. such that all $(d \times d)$–minors of $\mathcal{A}$ are either zer0, one, or negative one). Such a collection induces an infinite arrangement of hyperplanes

$$\{\boldsymbol{x} \in \mathbb{R}^d \ : \ \langle \boldsymbol{n}_i, \boldsymbol{x} \rangle = k\} \quad \text{for} \quad i = 1, \ldots, r \text{ and } k \in \mathbb{Z},$$

which subdivide $\mathbb{R}^d$ regularly into lattice polytopes. These subdivisions are referred to in the literature as lattice dicings [**49**].

A lattice polytope P whose collection of primitive facet normals forms a unimodular matrix is called *facet unimodular*. Every face of a facet unimodular polytope is also facet unimodular in its own lattice. The lattice dicing hyperplane arrangement slices P into dicing cells. This is called the *canonical subdivision* of a facet unimodular polytope. The canonical subdivision subdivides faces canonically.

THEOREM 2.4. *Suppose that $P \subset \mathbb{R}^d$ is a facet unimodular lattice polytope. Then:*

(1) *The canonical subdivision of P is regular, and all the cells are compressed.*
(2) *P has a regular unimodular triangulation.*

PROOF. The dicing cells have width one with respect to all their facets by construction. This proves part (1) except for regularity. Regularity of the canonical subdivision follows from considering weights given by the restriction of the following quadratic function to the lattice points in P:

$$f(\boldsymbol{x}) = \sum_{i=1}^{r} \langle \boldsymbol{n}_i, \boldsymbol{x} \rangle^2.$$

Since every lattice point in each cell of the canonical subdivision is contained in two consecutive hyperplanes in each direction $\boldsymbol{n}_i$, the corresponding summand of f is equal, on those lattice points, to an affine map. Therefore, f is affine on each cell. But any two adjacent cells contain lattice points in three hyperplanes for some direction, so f is not affine on the union.

For part (2) recall that any pulling refinement of a regular subdivision is regular. Since the cells in the canonical subdivision are compressed, a triangulation obtained by pulling will be regular and unimodular □

As a direct application of Theorem 2.4, flow polytopes (Section 3.3.1) as well as polytopes with facets in the root system of type **A** have regular unimodular triangulations (Section 3.1.1). This dicing method also shows that every dilation cP of a polytope P with a (regular) unimodular triangulation itself admits a (regular) unimodular triangulation (Theorem 4.8).

This approach to finding (regular) unimodular triangulations can be applied whenever there is a lattice dicing that cuts P into lattice polytopes. However in this more general case, the cells do not automatically have width one with respect to the facets given by P. This must be checked separately. Polytopes with facet normals in the root system of type **B** are an example where this approach was successful (Section 3.1.2).

2.1.4. Circuits. A *circuit* of a point configuration $\mathcal{A}$ is a minimal affine dependent subset C. A circuit C comes with a unique (up to a constant) affine dependence

$$\sum_{\boldsymbol{a}\in C} \lambda_{\boldsymbol{a}}\boldsymbol{a} = 0, \quad \sum_{\boldsymbol{a}\in C} \lambda_{\boldsymbol{a}} = 0.$$

It is well-known that a configuration that is itself a circuit, or that has a unique circuit, has exactly two triangulations. Having a unique circuit is equivalent to the configuration having exactly two points more than its dimension. The following is essentially Lemma 2.4.2 in [**44**].

LEMMA 2.5. *Let $\mathcal{A}$ be a configuration of $d+2$ points spanning a d-dimensional affine space. Let $\lambda \in \mathbb{R}^{\mathcal{A}}$ be its unique (up to a constant) affine dependence. Call*

$$\mathcal{A}^+ := \{\boldsymbol{a}\in\mathcal{A} : \lambda_{\boldsymbol{a}} > 0\}, \quad \mathcal{A}^0 := \{\boldsymbol{a}\in\mathcal{A} : \lambda_{\boldsymbol{a}} = 0\}, \quad \mathcal{A}^- := \{\boldsymbol{a}\in\mathcal{A} : \lambda_{\boldsymbol{a}} < 0\}.$$

Then, $\mathcal{A}$ has exactly two triangulations, namely:

$$\mathcal{T}^+ = \{F \subset \mathcal{A} : \mathcal{A}^+ \not\subset F\}, \qquad \mathcal{T}^- = \{F \subset \mathcal{A} : \mathcal{A}^- \not\subset F\},$$

Both triangulations are regular.

Put differently, $\mathcal{T}^+$ (resp. $\mathcal{T}^-$) has $\mathcal{A}^+$ (resp. $\mathcal{A}^-$) as its only minimal non-face. Observe that the points of $\mathcal{A}^0$ lie in every maximal simplex of both $\mathcal{T}^+$ and $\mathcal{T}^-$. This reflects the fact that for a $\boldsymbol{a}\in\mathcal{A}^0$, $\boldsymbol{a}$ is not in the affine span of $\mathcal{A}\setminus\boldsymbol{a}$.

It is easy to specify when these triangulations are flag and/or unimodular:

LEMMA 2.6. *Let $\mathcal{A}$ be a configuration of $d+2$ points spanning a d-dimensional affine space with its two triangulations $\mathcal{T}^+$ and $\mathcal{T}^-$.*

1. *$\mathcal{T}^+$ (resp. $\mathcal{T}^-$) is flag if, and only if, $|\mathcal{A}^+| \le 2$ (resp. $|\mathcal{A}^-| \le 2$).*
2. *Suppose $\mathcal{A}$ is a lattice point set and that λ is normalized to have integer entries with no common factor. Let $\Lambda_{\mathcal{A}}$ be the affine lattice generated by $\mathcal{A}$. Then, $\mathcal{T}^+$ (resp. $\mathcal{T}^-$) is unimodular* in $\Lambda_{\mathcal{A}}$ *if, and only if, all positive (resp. negative) coefficients in λ are equal to ± 1.*

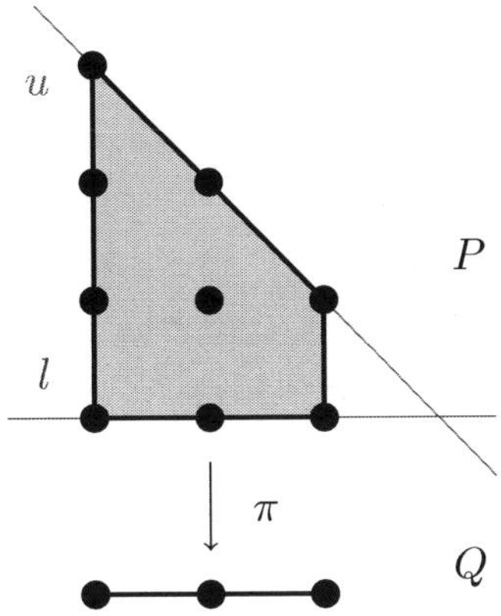

FIGURE 2.2. An example of the chimney construction. Here $l \equiv 0$ and $u(x) = 3 - x$ on $Q = [0, 2]$.

PROOF. For part (1), observe that $\mathcal{A}^+$ is the unique minimal non-face in $\mathcal{T}^+$. For part (2), observe that the coefficient $\lambda_{\boldsymbol{a}}$ of a point $\boldsymbol{a} \in \mathcal{A}$ equals ± 1 if, and only if, $\boldsymbol{a}$ is an integer affine combination of the rest of the points. Since the maximal simplices of $\mathcal{T}^+$ are precisely $\{\mathcal{A} \setminus \boldsymbol{a} : \boldsymbol{a} \in \mathcal{A}^+\}$, the result follows. □

In particular, for $\mathcal{T}^+$ to be quadratic we need $\mathcal{A}^+$ to have at most two elements and those elements have a coefficient of one in the dependence. Since $\sum_{\boldsymbol{a} \in \mathcal{A}} \lambda_a = 0$, $\mathcal{A}^-$ has also at most two elements and there are only the following two possibilities:

- $\mathcal{A}^+ = \{\boldsymbol{a}, \boldsymbol{b}\}$ and $\mathcal{A}^- = \{\boldsymbol{c}, \boldsymbol{d}\}$ with $\boldsymbol{a} + \boldsymbol{b} = \boldsymbol{c} + \boldsymbol{d}$, or
- $\mathcal{A}^+ = \{\boldsymbol{a}, \boldsymbol{b}\}$ and $\mathcal{A}^- = \{\boldsymbol{c}\}$ with $\boldsymbol{a} + \boldsymbol{b} = 2\boldsymbol{c}$.

The circuit consists of the four vertices of a parallelogram in the first case and of three collinear and equally spaced points in the second case.

Finally, let us mention that $\mathcal{T}^+$ (resp. $\mathcal{T}^-$) is the weak pulling of $\mathcal{A}$ from any $\boldsymbol{a} \in \mathcal{A}^-$ (resp. from any $\boldsymbol{a} \in \mathcal{A}^+$). It agrees with what strong pulling would give unless $\mathcal{A}^-$ (resp. $\mathcal{A}^+$) has a single element $\boldsymbol{a}$. (In this case, $\boldsymbol{a}$ is not a vertex of $\mathcal{A}$).

2.2. Push-forward subdivisions and pull-back subdivisions

In some cases the search for triangulations can be simplified via projection. This is done via push-forward and pull-back subdivisions.

2.2.1. Chimney polytopes and pull-back subdivisions. In this section we describe a method for recursively constructing unimodular triangulations of certain lattice polytopes. The process yields (regular) unimodular triangulations of *generalized prisms* over polytopes with a (regular) unimodular triangulation. In particular, this section extends results announced in [**60**].

We must first define chimney polytopes. Given a lattice polytope $Q \subset \mathbb{R}^d$, consider two integral linear functionals $\boldsymbol{l}$ and $\boldsymbol{u}$, such that $\boldsymbol{l} \leq \boldsymbol{u}$ along Q. We define the the *chimney polytope* associated to Q, $\boldsymbol{l}$ and $\boldsymbol{u}$ as

$$\operatorname{Chim}(Q, \boldsymbol{l}, \boldsymbol{u}) \ := \ \{(\boldsymbol{x}, y) \in \mathbb{R}^d \times \mathbb{R} \ \mid \ \boldsymbol{x} \in Q, \ \boldsymbol{l}(x) \leq y \leq \boldsymbol{u}(\boldsymbol{x})\}.$$

We call Q the *base* of the chimney. $\operatorname{Chim}(Q, \boldsymbol{l}, \boldsymbol{u})$ is itself a lattice polytope (see Figure 2.2).

We will show that a chimney polytope has a unimodular triangulation if its base has one. For this we introduce the general concept of a pull-back subdivision.

Given a lattice polytope P in $\mathbb{R}^d$ and a projection $\pi\colon \mathbb{R}^d \to \mathbb{R}^{d'}$, let $Q := \pi(P)$ and let $\mathcal{S}'$ be a subdivision of Q. The *pull-back subdivision* $\pi^*\mathcal{S}'$ of P is obtained from $\mathcal{S}'$ by intersecting P with the infinite prisms $\pi^{-1}(F)$ for each cell $F \in \mathcal{S}'$.

Observe that the cells in the pull-back subdivision may in principle not be lattice polytopes. A simple example is the projection of the triangle $\{(x, y)\colon 0 \leq 2y \leq x \leq 2\}$ to the segment $[0, 2]$. The pull-back of $\{[0, 1], [1, 2]\}$ produces cells with a non-integer vertex, $(1, 1/2)$. (Observe that this triangle is not a chimney polytope over the segment, because the functional $\boldsymbol{u}(x) = x/2$ is not integral).

We want to show that in the case when the pull-back is integral and the projection drops only one dimension, it can be refined to a triangulation with nice properties. To show that the construction preserves degree of the triangulation we need the following property:

LEMMA 2.7. *Let $\mathcal{T}$ be a simplicial ball whose dual graph is a tree. Then $\mathcal{T}$ is flag.*

PROOF. We use induction on the number of maximal simplices in $\mathcal{T}$.

Let F be a maximal simplex that corresponds to a leaf in the tree. F has a common facet with some other simplex in the triangulation, and a single vertex $\boldsymbol{a}$ not in that facet. Then, $\mathcal{T}' = \mathcal{T} \setminus F$ is also a triangulation whose dual graph is a tree, so we assume it is flag.

Let now N be a non-face of $\mathcal{T}$. If $\boldsymbol{a} \notin N$ then N is also a non-face in T' hence it has size two. If $\boldsymbol{a} \in N$ then pick any vertex $\boldsymbol{b} \in N \setminus F$ (which exists since $N \not\subset F$) and observe that $N' = \{\boldsymbol{a}, \boldsymbol{b}\}$ is a non-face. □

THEOREM 2.8. *Let $P \subset \mathbb{R}^d$ be a lattice polytope and let $\pi\colon \mathbb{R}^d \to \mathbb{R}^{d-1}$ be a projection such that $\pi(\mathbb{Z}^d) = \mathbb{Z}^{d-1}$. Let $\mathcal{T}$ be a unimodular triangulation of $Q := \pi(P)$ and suppose $\pi^*\mathcal{T}$ is integral, then*

- *Any full refinement $\mathcal{T}'$ of $\pi^*\mathcal{T}$ is a unimodular triangulation of P.*
- *$\mathcal{T}'$ does not have minimal non-faces of cardinality larger than those of $\mathcal{T}$.*
- *If $\mathcal{T}$ is regular, $\mathcal{T}'$ is regular as well.*

PROOF. To show that regularity can be preserved, recall that if $\mathcal{T}$ is regular, a full pulling refinement of $\pi^*\mathcal{T}$ will be regular as well.

For the unimodularity, it is enough to consider the chimneys $\pi^{-1}(G) \cap P$ for each simplex $G \in \mathcal{T}$ individually. They are equivalent to some $\operatorname{Chim}(\Delta^{d-1}, \mathbf{0}, \boldsymbol{u})$. Any d-simplex F in a full triangulation of $\operatorname{Chim}(\Delta^{d-1}, \mathbf{0}, \boldsymbol{u})$ has two vertices above one vertex of Δ^{d-1} and one vertex above every other vertex of Δ^{d-1}. Since F is part of a full triangulation $\mathcal{T}'$, the heights of the two vertices with the same projection differ by one. Hence, F is unimodular.

Let $N \subseteq P \cap \mathbb{Z}^d$ be a non-face of $\mathcal{T}'$. Then either $\pi(N)$ spans a face of $\mathcal{T}$ or not. If $\pi(N)$ is a non-face then N contains a non-face N' with $\pi(N') = \pi(N)$ on which π is injective.

If $\pi(N)$ is a face we can, again, restrict our attention to a single chimney of the form $\operatorname{Chim}(\Delta^{d-1}, \mathbf{0}, \boldsymbol{u})$. The dual graph in a triangulation of such a chimney is a path (cf. Figure 2.3) and Lemma 2.7 shows that N contains a non-face of cardinality two. □

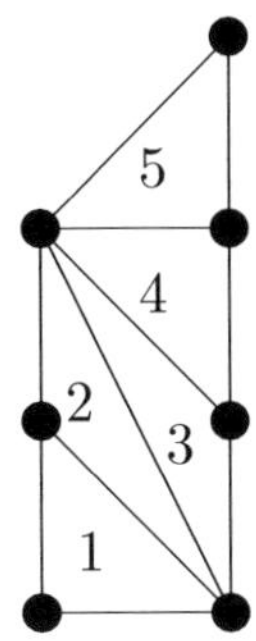

FIGURE 2.3. Ordering of maximal simplices in a simplex chimney.

This method of pull-back subdivisions and induction on dimension works nicely on the class of recursively defined polytopes known as Nakajima. A lattice polytope is a *Nakajima polytope* if it is a single lattice point or it is of the form $\mathrm{Chim}(Q, \mathbf{0}, \boldsymbol{u})$ for a Nakajima polytope Q. These are precisely those polytopes P for which the singularity U_P is a local complete intersection (see Section 1.2.4).

COROLLARY 2.9. *Every Nakajima polytope has a quadratic triangulation. (A triangulation which is regular, unimodular and flag.)*

PROOF. For a polytope $\mathrm{Chim}(Q, \boldsymbol{l}, \boldsymbol{u})$, the pull back of every lattice subdivision of Q is lattice. Hence, we can apply Theorem 2.8 recursively. □

2.2.2. Push-forward subdivision. To apply the chimney Theorem 2.8 in a case where P has more than one functional bounding P from below or from above, we need the subdivision of the projected polytope Q to respect the intersections of the multiple upper and lower facets. To this end we define the *push-forward* of a subdivision.

Given a subdivision $\mathcal{S}$ of a lattice polytope P in $\mathbb{R}^d$ and a projection $\pi\colon \mathbb{R}^d \to \mathbb{R}^{d'}$, the *push-forward subdivision* $\pi_*\mathcal{S}$ of $Q := \pi(P)$ is the common refinement of the projections of all faces of $\mathcal{S}$ (including low-dimensional faces).

The following theorem tells us under what conditions we can still apply Theorem 2.8.

THEOREM 2.10. *Let $P \subset \mathbb{R}^d$ be a lattice polytope and $\pi\colon \mathbb{R}^d \to \mathbb{R}^{d-1}$ be the projection which forgets the last coordinate. If $\mathcal{S}$ is a (regular) subdivision of P such that every cell F of $\mathcal{S}$ has a description*

$$F = \{(\boldsymbol{x}, y) \in \pi(F) \times \mathbb{R} \ : \ \boldsymbol{l}_i(\boldsymbol{x}) \leq y \leq \boldsymbol{u}_j(\boldsymbol{x}), \text{ for } 1 \leq i \leq r,\ 1 \leq j \leq s\}$$

with integral linear functionals $\boldsymbol{l}_1, \ldots, \boldsymbol{l}_r$ and $\boldsymbol{u}_1, \ldots, \boldsymbol{u}_s$ such that $\boldsymbol{l}_i \leq \boldsymbol{u}_j$ for $1 \leq i \leq r$, $1 \leq j \leq s$ along the lattice polytope Q, and that the push-forward $\pi_*\mathcal{S}$ of $\mathcal{S}$ to Q has a (regular) unimodular refinement, then $\mathcal{S}$ has a (regular) unimodular refinement. The degree of minimal non-faces will be preserved.

This theorem provides a heuristic for finding regular unimodular triangulations of lattice polytopes. Namely, given a lattice polytope P: search for unimodular transformations Φ of P such that $\Phi(P)$ has the above form; project to Q and check whether Q has a regular unimodular refinement of the push-forward subdivision; iterate. The push-forward and pull-back methods are implemented in an extension

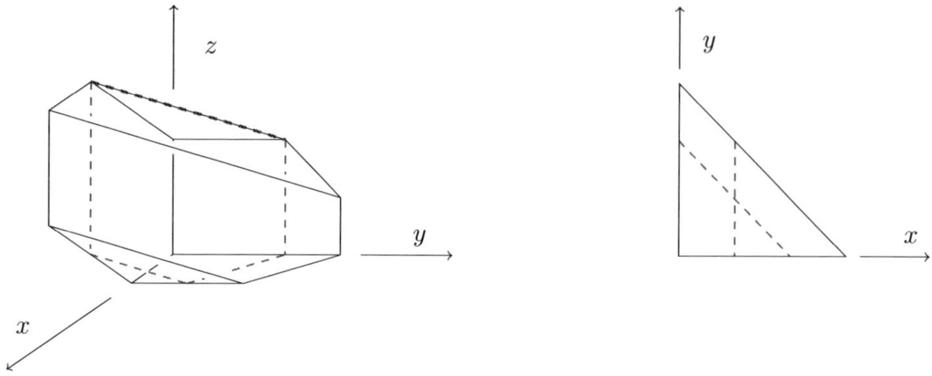

FIGURE 2.4. The projections P_{xyz} and P_{xy}

to polymake [**104**] and have been used for triangulating smooth reflexive polytopes (see Section 3.5.2).

Here is an example. Consider the following polytope given by eight inequalities in variables x, y, z, w.

$$\begin{array}{rcccl} 0 & \le & x & & \\ 0 & \le & y & \le & 3-x \\ 0 & \le & z & & \\ x-1 & \le & z & & \\ 0 & \le & w & \le & 2+x-z \\ & & w & \le & 4-y-z \end{array} \tag{2.1}$$

We have ordered the inequalities so that each variable is bounded above or below by integral linear functionals in the previous variables. We want to project P to x-y-z-space. This projection P_{xyz} has the representation (see Figure 2.4 on the left)

$$\begin{array}{rcccl} 0 & \le & x & & \\ 0 & \le & y & \le & 3-x \\ 0 & \le & z & \le & 2+x \\ x-1 & \le & z & \le & 4-y\,. \end{array}$$

Observe that P_{xyz} has facets $z \le 2+x$ and $z \le 4-y$ whose pull-backs are not facets of P. They are implied by the inequalities $0 \le w$ and $w \le 2+x-z$, respectively $w \le 4-y-z$.

The push-forward of the trivial subdivision of P divides P_{xyz} along the plane $x+y=2$, the projection of the ridge formed by the two upper bounds on w in (2.1),

$$\left.\begin{array}{rcccl} 0 & \le & w & \le & 2+x-z \\ & & w & \le & 4-y-z \end{array}\right\} x+y=2\,.$$

This is a lattice subdivision, as the intersection of this hyperplane with P_{xyz} is the convex hull of the lattice points $(1,1,0)$, $(0,2,0)$, $(0,2,2)$, $(2,0,4)$, and $(2,0,1)$.

We can project this again to obtain a subdivided polytope P_{xy} in the x-y-plane given by the inequalities $0 \le x$ and $0 \le y \le 3-x$ (see Figure 2.4 on the right). Any (regular and unimodular) triangulation of this subdivision can be used to construct a triangulation of P.

2.3. Joins and (Fiber) Products

2.3.1. Products. Let $\mathcal{T}$ and $\mathcal{T}'$ be subdivisions of P and P' respectively. Then

$$\mathcal{T} \times \mathcal{T}' := \{F \times F' \colon F \in \mathcal{T}, F' \in \mathcal{T}'\}$$

is the *product subdivision* of $P \times P'$. Unfortunately, the product of two triangulations is not a triangulation. It is a subdivision into products of simplices. The product of two unimodular simplices is totally unimodular (see [**132**, p.72, Ex.(9)], [**81**, p. 282], or [**44**, Section 6.2.2]; alternatively, think of it as the undirected edge polytope of a complete bipartite graph and apply Lemma 3.17(2)). In particular, all its triangulations are unimodular.

There is a particularly nice triangulation of a product of simplices $\Delta^d \times \Delta^{d'}$ (compare [**81**, p. 282], [**44**, Section 6.2.3]). To define it, order the vertices of the factors $\boldsymbol{a}_0 \prec \ldots \prec \boldsymbol{a}_d$ and $\boldsymbol{a}'_0 \prec \ldots \prec \boldsymbol{a}'_{d'}$. This induces a componentwise partial order on the vertices of $\Delta^d \times \Delta^{d'}$. The family of totally ordered subsets yields a quadratic triangulation, called the *staircase triangulation.* One geometric way to construct this triangulation is to pull the vertices of $\Delta^d \times \Delta^{d'}$ in the lexicographic order.

PROPOSITION 2.11. *Let P and P' be lattice polytopes. If both admit regular unimodular triangulations $\mathcal{T}$ and $\mathcal{T}'$, then so does $P \times P'$.*

The set of minimal non–faces consists of lifts of minimal non–faces from P and P' together with non–faces of cardinality two.

PROOF. If $\mathcal{T}$ and $\mathcal{T}'$ are regular, then the subdivision $\mathcal{T} \times \mathcal{T}'$ of $P \times P'$ into products of unimodular simplices is regular, and any triangulation that refines it is unimodular. In order to control the non–faces, order the lattice points $\boldsymbol{p}_1 \prec \ldots \prec \boldsymbol{p}_r$ in P and $\boldsymbol{p}'_1 \prec \ldots \prec \boldsymbol{p}'_s$ in P'. Then pull the lattice points $(\boldsymbol{p}_i, \boldsymbol{p}'_j)$ in $P \times P'$ lexicographically.

Consider a non–face N. If both its projections to P or P' are faces, N is a non–face in a staircase triangulation. □

Proposition 2.11 can be extended to non-regular triangulations. This extension is at the heart of the counter-examples constructed by Santos in [**114, 115**] (see also [**44**, Ch. 7]) which we will return to in Section 3.6. The main idea is that if we do not care about the regularity of the unimodular triangulation of $P \times P'$ we do not need to refine $\mathcal{T} \times \mathcal{T}'$ by pulling vertices. Any refinement of the individual cells of $\mathcal{T} \times \mathcal{T}'$ is unimodular, and the only concern is that the different refinements agree on common faces. Using staircase refinements of each product of simplices will still accomplish this. It does not require a globally defined ordering of all the vertices of $\mathcal{T}$ and of $\mathcal{T}'$, but only a local ordering of the vertices in each individual simplex. These local orderings can be represented via an acyclic orientation of the 1-skeleton of each simplex, as follows. Let $\mathcal{T}$ be a triangulation of a point configuration. A *locally acyclic orientation* of the one-skeleton of $\mathcal{T}$ (or a locally acyclic orientation of $\mathcal{T}$) is an assignment of a direction to each edge such that no simplex contains a directed cycle (equivalently, no triangle in $\mathcal{T}$ is a directed three-cycle).

PROPOSITION 2.12. *Let $\mathcal{T}_1$ and $\mathcal{T}_2$ be triangulations of P_1 and P_2 with locally acyclic orientations. Refining each product of simplices in $\mathcal{T}_1 \times \mathcal{T}_2$ in the staircase manner indicated by the orientations produces a triangulation $\mathcal{T}$ of $P_1 \times P_2$ with the following properties:*

(1) $\mathcal{T}$ *is unimodular if and only if* $\mathcal{T}_1$ *and* $\mathcal{T}_2$ *are both unimodular.*
(2) $\mathcal{T}$ *is regular if and only if* $\mathcal{T}_1$ *and* $\mathcal{T}_2$ *are both regular and the orientations are globally acyclic.*
(3) $\mathcal{T}$ *is flag if and only if* $\mathcal{T}_1$ *and* $\mathcal{T}_2$ *are both flag.*

The triangulation of Proposition 2.12 is called the *staircase refinement* of $\mathcal{T}_1 \times \mathcal{T}_2$ with respect to the corresponding locally acyclic orientations.

PROOF. We made it clear above that the triangulation is well-defined. For (1) the refinement is unimodular if the factors are unimodular since every triangulation of the product of two simplices is unimodular. Conversely, if a simplex in one of the factors is not unimodular, then the staircase refinement does not give product cells using that simplex unimodular refinements. For (2), observe that the restriction of $\mathcal{T}$ to every $\{\boldsymbol{a}_1\} \times P_2$ or $P_1 \times \{\boldsymbol{a}_2\}$ for vertices $\boldsymbol{a}_1$ of P_1 or $\boldsymbol{a}_2$ of P_2 is affinely isomorphic to $\mathcal{T}_2$ or $\mathcal{T}_1$, respectively. So, both must be regular for $\mathcal{T}$ to be regular. If the locally acyclic orientation of, say, $\mathcal{T}_1$, has a cycle $\boldsymbol{a}_0, \boldsymbol{a}_1, \ldots, \boldsymbol{a}_k = \boldsymbol{a}_0$ then for every (oriented) edge $\boldsymbol{b}_1\boldsymbol{b}_2$ in $\mathcal{T}_2$ the edges $(\boldsymbol{a}_i, \boldsymbol{b}_1)(\boldsymbol{a}_{i+1}, \boldsymbol{b}_2)$, $i = 0, \ldots, k$ are in $\mathcal{T}$, which implies $\mathcal{T}$ is not regular. Indeed, the circuit

$$(\boldsymbol{a}_i, \boldsymbol{b}_1) + (\boldsymbol{a}_{i+1}, \boldsymbol{b}_2) = (\boldsymbol{a}_i, \boldsymbol{b}_2)(\boldsymbol{a}_{i+1}, \boldsymbol{b}_1)$$

implies that if the edge $(\boldsymbol{a}_i, \boldsymbol{b}_1)(\boldsymbol{a}_{i+1}, \boldsymbol{b}_2)$ appears the following inequality must hold for the weight vector $\boldsymbol{\omega}$:

$$\boldsymbol{\omega}(\boldsymbol{a}_i, \boldsymbol{b}_1) + \boldsymbol{\omega}(\boldsymbol{a}_{i+1}, \boldsymbol{b}_2) < \boldsymbol{\omega}(\boldsymbol{a}_i, \boldsymbol{b}_2)\boldsymbol{\omega}(\boldsymbol{a}_{i+1}, \boldsymbol{b}_1).$$

Summing this over all i yields the impossible equation

$$\sum_i \boldsymbol{\omega}(\boldsymbol{a}_i, \boldsymbol{b}_1) + \sum_i \boldsymbol{\omega}(\boldsymbol{a}_i, \boldsymbol{b}_2) < \sum_i \boldsymbol{\omega}(\boldsymbol{a}_i, \boldsymbol{b}_2) \sum_i \boldsymbol{\omega}(\boldsymbol{a}_i, \boldsymbol{b}_1).$$

This proves it is necessary for $\mathcal{T}_1$ and $\mathcal{T}_2$ to be regular. For sufficiency, note that if $\mathcal{T}_1$ and $\mathcal{T}_2$ are regular, $\mathcal{T}_1 \times \mathcal{T}_2$ is a regular subdivision. Now if the orientations of $\mathcal{T}_1$ and $\mathcal{T}_2$'s 1-skeletons are globally acyclic, they can be extended to give total orderings on the vertices of $\mathcal{T}_1$ and $\mathcal{T}_2$, and our staircase refinement can be obtained by pulling the vertices of $\mathcal{T}_1 \times \mathcal{T}_2$ with respect to the lexicographic ordering, which yields a regular triangulation.

For (3), flagness follows from the characterization of minimal non-faces of the staircase refinement stated in Proposition 2.11. □

2.3.2. Joins. Let P and P' be polytopes of dimension d and d', and $\mathbf{0}_k$ the origin in $\mathbb{R}^k$. The join $P \star P'$ of P and P' is the convex hull of

$$P \times \{\mathbf{0}_{d'}\} \times \{0\} \quad \cup \quad \{\mathbf{0}_d\} \times P' \times \{1\}.$$

This gives a $(d + d' + 1)$-dimensional polytope. The join of two simplices, $\Delta^r \star \Delta^{r'}$ is a simplex $\Delta^{r+r'+1}$. Any subdivisions $\mathcal{S}$ and $\mathcal{S}'$ of P and P' lift to a subdivision $\mathcal{T}$ of $P \star P'$ by taking all joins of cells in $\mathcal{S}$ and $\mathcal{S}'$, and every subdivision of $P \star P'$ can be obtained this way. In particular, if $\mathcal{S}$ and $\mathcal{S}'$ are (regular, unimodular, flag) triangulations of P and P', then $\mathcal{T}$ is a (regular, unimodular, flag) triangulation of $P \star P'$.

The toric ring of the join is the tensor product of the components: $R_{P \star P'} = R_P \otimes R_{P'}$ (compare Section 1.2.3).

The facets of $P \star P'$ are joins of P with facets of P' and joins of P' with facets of P. Hence $P \star P'$ is compressed if and only if both P and P' are compressed.

The join can be defined for more than two factors in a similar way, and is associative. Just as $P \star P'$ has a canonical projection to Δ^1 — the last coordinate, $P_0 \star \ldots \star P_r$ has a canonical projection to Δ^r.

2.3.3. Fiber Products. Suppose two lattice polytopes project linearly, respecting their lattices, to the same lattice polytope: $P \overset{\pi}{\to} Q \overset{\pi'}{\leftarrow} P'$. Then the *polyhedral fiber product*, also known as the *multigraded Segre product* $P \times_Q P'$ is the polytope

$$\{(\boldsymbol{p}, \boldsymbol{p}') \in P \times P' \ : \ \pi(\boldsymbol{p}) = \pi'(\boldsymbol{p}')\} \ .$$

This construction was first used by Buczyńska and Wisniewski in the study of statistical models of binary symmetric phylogenetic trees [**35**]. A closely related *toric fiber product*

$$\operatorname{conv}\left\{(\boldsymbol{p}, \boldsymbol{p}') \in (P \cap \mathbb{Z}^d) \times (P' \cap \mathbb{Z}^{d'}) \ : \ \pi(\boldsymbol{p}) = \pi'(\boldsymbol{p}')\right\}$$

was defined by Sullivant [**135**].

Under the assumptions of the following theorem (which includes the phylogenetic case) $P \times_Q P'$ is a lattice polytope, so the two notions agree.

THEOREM 2.13. *Let $P \overset{\pi}{\to} Q \overset{\pi'}{\leftarrow} P'$ be lattice preserving projections. If Q admits a unimodular triangulation $\mathcal{T}$, and P and P' have unimodular triangulations $\mathcal{S}$ and $\mathcal{S}'$, which refine the pull-back subdivisions $\pi^*\mathcal{T}$ and $\pi'^*\mathcal{T}$ respectively, then $P \times_Q P'$ admits a unimodular triangulation.*

Further, regularity and the degree of minimal non-faces can be preserved.

Before proving this theorem, let us state as a corollary a slight generalization of a result of Sullivant [**135**, Cor. 15] for the case where Q is a unimodular simplex. In this case, the pull-back subdivision is trivial.

COROLLARY 2.14. *If $P \overset{\pi}{\to} \Delta^r \overset{\pi'}{\leftarrow} P'$ are lattice preserving projections such that both P and P' admit unimodular triangulations, then $P \times_{\Delta^r} P'$ admits a unimodular triangulation.*

Regularity and degree of minimal non-faces can be preserved.

A lattice polytope projecting to a unimodular simplex is known in the literature as a Cayley sum of the fibers of the simplex vertices [**56**, Ch.9, eq.(1.2)]. The fiber product in the above corollary is the Cayley sum of the products of the fibers.

The proof of Theorem 2.13 requires the following lemma.

LEMMA 2.15. *Given lattice preserving projections $\Delta^d \overset{\pi}{\to} \Delta^r \overset{\pi'}{\leftarrow} \Delta^{d'}$, the fiber product $\Delta^d \times_{\Delta^r} \Delta^{d'}$ is a lattice polytope.*

PROOF. For $\nu = 1, \ldots, r+1$ let $I_\nu := \{\ i \ : \ \pi(\boldsymbol{e}_i) = \boldsymbol{e}_\nu\}$ and $I'_\nu := \{\ j \ : \ \pi'(\boldsymbol{e}_j) = \boldsymbol{e}_\nu\}$. With this notation, the fiber product has the inequality description

$$(\boldsymbol{p}, \boldsymbol{p}') \geq \boldsymbol{0} \ ,$$

$$\sum_{i=1}^{d+1} p_i = \sum_{j=1}^{d'+1} p'_j = 1 \ ,$$

$$\sum_{i \in I_\nu} p_i = \sum_{j \in I'_\nu} p'_j \quad \text{for} \quad 1 \leq \nu \leq r+1 \ .$$

The equation matrix is (after omission of repeated columns) of the form

$$\begin{pmatrix} 1\cdots 1 & 0\cdots 0 \\ 0\cdots 0 & 1\cdots 1 \\ \mathrm{id}_{r+1} & -\mathrm{id}_{r+1} \end{pmatrix}$$

which is a totally unimodular matrix. □

PROOF OF THEOREM 2.13. Let $\mathcal{T}$, $\mathcal{S}$ and $\mathcal{S}'$ be as in the statement. $\mathcal{S}$ and $\mathcal{S}'$ give a subdivision $\mathcal{S}\times\mathcal{S}'$ of $P\times P'$ into products of unimodular simplices. We claim that intersecting $\mathcal{S}\times\mathcal{S}'$ with $P\times_Q P'$ gives a lattice subdivision of $P\times_Q P'$. Indeed, consider a cell in this subdivision $(F\times F')\cap(P\times_Q P')$ a cell in this subdivision, for some unimodular simplices F and F'. Since π and π' map simplices of $\mathcal{S}$ and $\mathcal{S}'$ to simplices in $\mathcal{T}$, πF and $\pi F'$ are simplices in $\mathcal{T}$ and, in fact, (assuming $F\times F'$ is the minimal product of simplices containing our cell) they are the same simplex G. Then

$$(F\times F')\cap(P\times_Q P') = F\times_G F'$$

which, by Lemma 2.15, is a lattice polytope.

Thus, we have a lattice regular subdivision of $P\times_Q P'$ into totally unimodular cells. (For regularity, observe that the intersection of a regular subdivision with an affine subspace is regular). Any refinement of it into a triangulation of $P\times_Q P'$ is unimodular. If the refinement is done, for example, pulling all the lattice points in $P\times_Q P'$ lexicographically as in the proof of Proposition 2.11, it will also be regular. The non–face statement follows as in Proposition 2.11. □

Note that Theorem 2.13 is true for more than two factors, by induction, since the triangulation obtained refines the pull-back of $P\times_Q P'\to Q$.

2.3.4. Semidirect products. Motivated by a construction in algebraic statistics, Aoki, Hibi, Ohsugi and Takemura introduced *nested configurations.* Given lattice polytopes $Q\subseteq k\Delta^d$ and $P_i\subset\mathbb{R}^{d_i}$ for $i=1,\dots,d+1$, the *nested polytope* $\mathrm{NP}(Q;P_1,\dots,P_{d+1})$ is the convex hull, in $\mathbb{R}^{d+1}\times\prod\mathbb{R}^{d_i}$ of the polytopes $\{\boldsymbol{a}\}\times\prod a_iP_i$, where $\boldsymbol{a}$ runs over the vertices of Q. Here and in what follows, $\Delta^d=\{(x_1,\dots,x_{d+1}) : \boldsymbol{x}\geq 0, \sum x_i=1\}$ denotes the homogeneous unimodular d-simplex in $\mathbb{R}^{d+1}$.

The following is an equivalent definition in terms of joins.

$$\mathrm{NP}(Q;P_1,\dots,P_{d+1}) = k\cdot(P_1\star\cdots\star P_{d+1})\cap\pi^{-1}(Q),$$

where $\pi:\mathbb{R}^{d+1}\times\prod\mathbb{R}^{d_i}\to\mathbb{R}^{d+1}$ is the natural projection.

In [**103**], Hibi and Ohsugi show that many properties of the input polytopes can be inherited by nested polytopes. These include normality and the existence of regular unimodular triangulations as well as degrees of Gröbner bases (or generators). Their proof uses the algebraic-geometric machinery from section 2.4, so no statements about non-regular triangulations can be concluded from it. Here we offer a purely combinatorial proof. (The definition in [**4**] takes more general configurations as input, but if one is only interested in the normal case, no generality is lost by taking our definition.)

We first introduce the following alternative way of looking at nested configurations. Given lattice polytopes $Q\subset\mathbb{R}^d$ and $P_i\subset\mathbb{R}^{d_i}$ for $i=1,\dots,n$ and an integer affine map, $\phi : \mathbb{Z}^d \to \mathbb{Z}^n$, that is nonnegative on Q, the *semidirect product* of Q and the tuple $(P_1, \dots, P_n)$ along the map ϕ is defined as

$$Q \ltimes_\phi (P_1, \dots, P_n) := \operatorname{conv}_{\boldsymbol{a} \in Q} \left(\{\boldsymbol{a}\} \times \prod \phi_i(\boldsymbol{a}) P_i \right),$$

where $(\phi_1, \dots, \phi_n)$ are the coordinates of ϕ.

If $n = d+1$, $Q \subset k\Delta^d$, and ϕ is the identity map, we recover the definition of nested configuration. Conversely, every semidirect product can be rewritten as a nested configuration as follows. If ϕ is injective, unimodular (meaning that $\phi(\mathbb{Z}^d) = \operatorname{aff}(\phi(\mathbb{Z}^d)) \cap \mathbb{Z}^n)$) and homogeneous (meaning that $\phi(\mathbb{Z}^d) \subset \{\sum x_i = k\}$ for some $k \in \mathbb{N}$) then

$$Q \ltimes_\phi (P_1, \dots, P_n) \cong \operatorname{NP}(\phi(Q); P_1, \dots, P_n).$$

If ϕ is not injective, not unimodular, or not homogeneous, consider the modified map $\widetilde{\phi} = (\phi, \operatorname{Id}, f) : \mathbb{R}^d \to \mathbb{R}^n \times \mathbb{R}^d \times \mathbb{R}$, where $f(x) = k - \sum_i \phi(x)_i - \sum_j x_j$ for a sufficiently large k, take P_i to be a single point for all $i > n$ and observe that

$$Q \ltimes_\phi (P_1, \dots, P_n) \cong \operatorname{NP}(\widetilde{\phi}(Q); P_1, \dots, P_n, \mathrm{pt}, \dots, \mathrm{pt}).$$

So, semidirect products are really an equivalent operation to nested configurations. But we find them conceptually easier to handle. They generalize the following constructions:

- $\Delta^d \ltimes_{\mathrm{Id}} (P_0, \dots, P_d)$ is the join of $P_0, \dots, P_d$,
- $\{\mathrm{pt}\} \ltimes_1 (P_0, \dots, P_d)$ is the product of $P_0, \dots, P_d$, as is $P_0 \ltimes_1 (P_1, \dots, P_d)$. In both cases 1 denotes the constant map with image $(1, \dots, 1)$.
- $\{\mathrm{pt}\} \ltimes_k (P)$ is the k-th dilation of P and $\{\mathrm{pt}\} \ltimes_{(k_1,\dots,k_d)} (P_1, \dots, P_d)$ is the product $\prod k_i P_i$.
- The chimney $\operatorname{Chim}(Q, \boldsymbol{l}, \boldsymbol{u})$ associated to two integer functionals $\boldsymbol{l} \le \boldsymbol{u}$ on Q is equivalent to the semidirect product $Q \ltimes_{\boldsymbol{u}-\boldsymbol{l}} I$, where I is a unimodular segment. In particular, a Nakajima polytope is one that can be obtained as
$$(\dots(\{\mathrm{pt} \ltimes_{\phi_1} I) \dots) \ltimes_{\phi_d} I,$$
for certain choice of functionals ϕ_i.

There are two ways of relating any semidirect product with several factors to semidirect products with only two factors at a time. One is as a special case of fiber products, taking into account that every semidirect product comes with a canonical projection $Q \ltimes_\phi (P_1, \dots, P_n) \to Q$. The other is a special associativity property:

LEMMA 2.16. *As in the definition of the semidirect product, let $\phi_1, \dots, \phi_n$ denote the coordinates of $\phi : \mathbb{R}^d \to \mathbb{R}^n$. Then:*

$$\begin{aligned} Q \ltimes_\phi (P_1, \dots, P_d) &= (Q \ltimes_{\phi_1} P_1) \times_Q \cdots \times_Q (Q \ltimes_{\phi_n} P_n) \\ &= (\dots(Q \ltimes_{\widetilde{\phi}_1} P_1) \ltimes_{\widetilde{\phi}_2} \cdots \ltimes_{\widetilde{\phi}_d} P_d), \end{aligned}$$

where $\widetilde{\phi}_i$ denotes the composition of the natural projection $\mathbb{Z}^d \times \mathbb{Z}^{d_1 + \dots + d_{i-1}} \to \mathbb{Z}^d$ with $\phi_i : Z^d \to \mathbb{Z}$. □

We are going to prove that

THEOREM 2.17. *Suppose Q and P admit unimodular triangulations $\mathcal{S}$ and $\mathcal{T}$. Then, the semidirect product $Q \ltimes_\phi P$ admits a unimodular triangulation that refines the pull-back $\pi^*\mathcal{S}$ of $\mathcal{S}$ by the projection $\pi : Q \ltimes_\phi P \to Q$.*

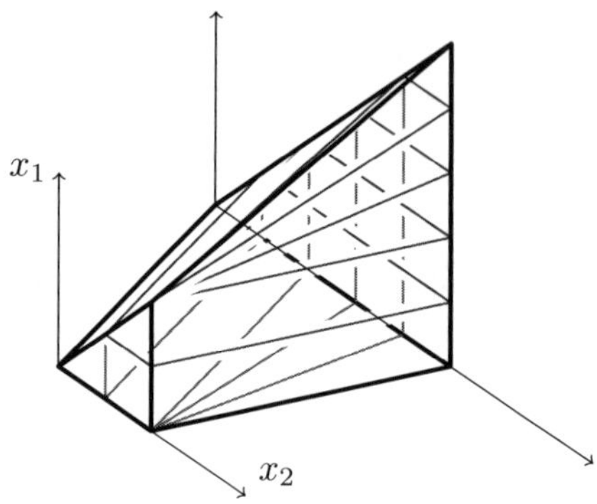

FIGURE 2.5. The hyperplanes $H(1,b)$ and $H(2,b)$ for $d=1$, $e=2$, and $\phi(\boldsymbol{q}_1)=2$, $\phi(\boldsymbol{q}_2)=5$

COROLLARY 2.18. *If $Q, P_1, \ldots,$ and P_n admit unimodular triangulations, then every semidirect product $Q \ltimes_\phi (P_1, \ldots, P_n)$ admits one too.*

PROOF. Lemma 2.16 gives two different ways to derive the corollary from Theorem 2.17. One is by associativity, using induction on n. The other is via the relation to fiber products, using Theorem 2.13. □

The key step for the proof of Theorem 2.17 is to look at the case where both Q and P are unimodular simplices. So, let $Q \subset \mathbb{R}^d$ be a unimodular d-simplex and $P = \operatorname{conv}\{\boldsymbol{p}_0, \ldots, \boldsymbol{p}_e\} \subset \mathbb{R}^e$ a unimodular e-simplex. As implied by our notation, the vertices of P are considered with a given specific order, which will be important both for the construction on this particular simplex and for gluing the constructions between simplices. Let $f_j : P \to \mathbb{R}$ be the affine functional that takes the value zero in $\boldsymbol{p}_0, \ldots, \boldsymbol{p}_{e-i}$ and one in $\boldsymbol{p}_{e-j+1}, \ldots, \boldsymbol{p}_e$. Put differently:

$$P = \{\boldsymbol{x} \in \mathbb{R}^e \ : \ 0 \le f_1(\boldsymbol{x}) \le \cdots \le f_e(\boldsymbol{x}) \le 1\}.$$

(If P is the *standard ordered e-simplex* $\{\boldsymbol{x} : 0 \le x_1 \le \cdots \le x_e \le 1\}$ then f_j is the j-th coordinate).

Let $\phi : Q \to \mathbb{R}$ be the affine functional in the definition of semidirect product. Then,

$$Q \ltimes_\phi P = \left\{ \begin{pmatrix} \boldsymbol{y} \\ \boldsymbol{x} \end{pmatrix} \in Q \times \mathbb{R}^e \ : \ 0 \le f_1(\boldsymbol{x}) \le \cdots \le f_e(\boldsymbol{x}) \le \phi(\boldsymbol{y}) \right\}.$$

In this setting we can define the canonical slicing of $Q \ltimes_\phi P$. For each $b \in \mathbb{R}$ let $\phi_{\le b} : Q \to \mathbb{R}$ be the (unique) affine functional with $\phi_{\le b}(\boldsymbol{q}) = \min(\phi(\boldsymbol{q}), b)$ on each vertex $\boldsymbol{q}$ of Q, and for each $b \in \mathbb{N}$ and each $j \in [e]$ consider the hyperplane

$$H(j,b) := \left\{ \begin{pmatrix} \boldsymbol{y} \\ \boldsymbol{x} \end{pmatrix} \in \mathbb{R}^d \times \mathbb{R}^e \ : \ f_j(\boldsymbol{x}) = \phi_{\le b}(\boldsymbol{y}) \right\}.$$

The *canonical slicing* of $Q \ltimes_\phi P$ is the polyhedral subdivision obtained slicing $Q \ltimes_\phi P$ by all these hyperplanes.

Figure 2.5 shows the canonical slicing in the case $d=1$, $e=2$, with ϕ taking the values two and five on the vertices of the segment Q.

LEMMA 2.19. *Let Q and $P = \operatorname{conv}\{\boldsymbol{p}_0, \dots, \boldsymbol{p}_e\}$ be unimodular simplices, with the vertices of P given in a specified ordering, and let $\phi : Q \to \mathbb{R}$ be a nonnegative integer affine function on Q. Then:*

(1) *The canonical slicing of every face of $Q \ltimes_\phi P$ coincides with the restriction to that face of the canonical slicing of $Q \ltimes_\phi P$.*
(2) *The canonical slicing is a lattice subdivision (all vertices are integer).*
(3) *All cells in the canonical slicing are compressed.*

PROOF. Part (1) is trivial.

For part (2), let $\binom{\bar{\boldsymbol{y}}}{\bar{\boldsymbol{x}}} \in Q \ltimes_\phi P$ be a vertex of the canonical slicing. By part (1) there is no loss of generality in assuming that $\binom{\bar{\boldsymbol{y}}}{\bar{\boldsymbol{x}}}$ lies in the interior of $Q \ltimes_\phi P$, in particular $\bar{\boldsymbol{y}}$ lies in the interior of Q. For such an $\bar{\boldsymbol{y}}$ the function $b \mapsto \phi_{\le b}(\bar{\boldsymbol{y}})$ is (continuous and) strictly increasing in the range $b \le \max_{\boldsymbol{q}}(\phi(\boldsymbol{q}))$. Hence $\binom{\bar{\boldsymbol{y}}}{\bar{\boldsymbol{x}}}$ lies in at most one hyperplane $H(j, b)$ for each $j \in [e]$. In order for these hyperplanes to define a vertex, we need to have (at least) $d+e$ of them, so that the only possibility is $d = 0$. But when $d = 0$, $Q \ltimes_\phi P$ is just the k-th dilation of P, where k is the value taken by ϕ in the (unique) point of Q. The hyperplanes $H(j, b)$ are of the form "f_j equals a constant", and the facet-defining hyperplanes of kP are of the form $f_{j+1} - f_j = 0$. Together they form a totally unimodular system of possible facet normals (in the basis consisting of the f_j's, which is itself unimodular), so the slicing they produce can only have integer vertices.

The inductive argument above implies that every vertex of the canonical slicing lies in one of the fibers $\{\boldsymbol{q}\} \times \phi(\boldsymbol{q})P$ where $\boldsymbol{q}$ is a vertex of Q, which will be useful in the last part of the proof.

For part (3), we consider the three possible types of facets separately.

- Those contained in facets $F \ltimes_\phi P$, where F is a facet of Q, have width one since Q is unimodular and $Q \ltimes_\phi P$ projects to it.
- For, those defined by a hyperplane $H(j, b)$, observe that cells incident to that hyperplane are contained between the hyperplanes $H(j, b-1)$ and $H(j, b+1)$. For every point $\binom{\bar{\boldsymbol{y}}}{\bar{\boldsymbol{x}}}$ in those cells we have
$$\phi_{\le b}(\bar{\boldsymbol{y}}) - 1 \le \phi_{\le b-1}(\bar{\boldsymbol{y}}) \le f_j(\bar{\boldsymbol{x}}) \le \phi_{\le b+1}(\bar{\boldsymbol{y}}) \le \phi_{\le b}(\bar{\boldsymbol{y}}) + 1,$$
which proves they have width one.
- For, those contained in facets $Q \ltimes_\phi F_j$ of $Q \ltimes_\phi P$, where F_j is a facet of P, recall that F_j is defined by the equation $f_j(\boldsymbol{x}) = f_{j-1}(\boldsymbol{x})$, $j = 1, \dots, e+1$ (with the convention $f_0 \equiv 0$ and $f_{e+1} \equiv 1$). The facet $Q \ltimes_\phi F_j$ lies in the hyperplane (in $\mathbb{R}^d \times \mathbb{R}^e$ defined by the same equation (in the $\mathbb{R}^e$ variables), except for the $j = e+1$ case, where the equation defining $Q \ltimes_\phi F_{e+1}$ is $f_{e+1}(\boldsymbol{x}) = \phi(\boldsymbol{y})$. All vertices of the canonical slicing are in fibers over the vertices of Q. Since the canonical slicing restricted to these fibers is a dicing by a system of totally unimodular vectors (the same system of vectors as in the proof of part (2)), the cells in each individual fiber have width one. This implies that the cells in the whole slicing also have width one with respect to these functionals.

□

PROOF OF THEOREM 2.17. Suppose Q, P admit unimodular triangulations $\mathcal{S}$, $\mathcal{T}$ respectively. We then have that the cells

$$B \ltimes_\phi C,$$

for all the simplices $B \in \mathcal{S}$ and $C \in \mathcal{T}$ form a subdivision of $Q \ltimes_\phi P$. (To show this, if ϕ is strictly positive on Q observe that $Q \ltimes_\phi P$ is projectively equivalent to $Q \times P$. If ϕ is zero on some face of Q then the corresponding face of $Q \times P$ collapses to lower dimension in $Q \ltimes_\phi P$, but the result is still true).

Now, Lemma 2.19 tells us how to unimodularly subdivide each $B \ltimes_\phi C$ into compressed lattice polytopes, and the canonical nature of these subdivisions guarantees that they agree on common faces. Any triangulation that refines the subdivision obtained this way (e.g., by pulling all vertices) is unimodular. □

2.4. Toric Gröbner Bases

The toric dictionary translates between the discrete geometry of lattice points in polytopes and the algebraic geometry of toric varieties. We explore the translations involving unimodular triangulations.

2.4.1. Unimodular triangulations and Gröbner bases. Let $\Bbbk$ be a field, and $\mathcal{A} := (P \times \{1\}) \cap \mathbb{Z}^{d+1}$ denote the homogenized set of lattice points in P. Consider the polynomial ring $S := \Bbbk[x_{\boldsymbol{a}} : \boldsymbol{a} \in \mathcal{A}]$ with one variable for each lattice point. There is a canonical ring homomorphism ϕ_P to the Laurent polynomial ring $\Bbbk[t_1^{\pm 1}, \ldots, t_d^{\pm 1}, t_{d+1}]$ mapping each variable to the corresponding (homogenized) $\boldsymbol{t}$-monomial: $\phi_P(x_{\boldsymbol{a}}) = \boldsymbol{t}^{\boldsymbol{a}} = t_1^{a_1} \cdot \ldots \cdot t_d^{a_d} \cdot t_{d+1}$. The *toric ideal* $\mathcal{I}_P := \ker \phi_P$ is spanned, as a $\Bbbk$-vector space, by the set

$$\left\{ \boldsymbol{x}^{\boldsymbol{u}} - \boldsymbol{x}^{\boldsymbol{v}} \;:\; \boldsymbol{u}, \boldsymbol{v} \in \mathbb{Z}_{\geq 0}^{\mathcal{A}},\ \sum_{\boldsymbol{a} \in \mathcal{A}} u_{\boldsymbol{a}} \boldsymbol{a} = \sum_{\boldsymbol{a} \in \mathcal{A}} v_{\boldsymbol{a}} \boldsymbol{a} \right\}, \tag{2.2}$$

where $\boldsymbol{x}^{\boldsymbol{u}} = \prod_{\boldsymbol{a} \in \mathcal{A}} x_{\boldsymbol{a}}^{u_{\boldsymbol{a}}}$. It thus encodes the affine dependencies among the lattice points in P [**132**, Lemma 4.1].

The tie between Gröbner bases of $\mathcal{I}_P$ and regular triangulations of P is established via two different interpretations of a generic[1] weight vector $\boldsymbol{\omega} \in \mathbb{R}^{\mathcal{A}}$. On the lattice polytope side, $\boldsymbol{\omega}$ induces a regular triangulation $\mathcal{T}_{\boldsymbol{\omega}}$ of P as explained at the end of Section 1.1. On the algebraic side, such an $\boldsymbol{\omega}$ induces an ordering of the monomials in S via $\boldsymbol{x}^{\boldsymbol{m}} \prec \boldsymbol{x}^{\boldsymbol{n}} \;:\Longleftrightarrow\; \langle \boldsymbol{\omega}, \boldsymbol{m} \rangle < \langle \boldsymbol{\omega}, \boldsymbol{n} \rangle$. For a polynomial $f \in S$ the *leading term* $\mathrm{in}_{\boldsymbol{\omega}} f$ is the biggest monomial in this ordering which has a non-zero coefficient in f, and the *initial ideal* $\mathrm{in}_{\boldsymbol{\omega}} I := \langle \mathrm{in}_{\boldsymbol{\omega}} f \;:\; f \in I \rangle$ of an ideal I collects all the leading terms of polynomials in I. A collection of polynomials in I whose leading terms generate the initial ideal is called a *Gröbner basis* of I (with respect to $\boldsymbol{\omega}$). If $\boldsymbol{\omega}$ is not generic, we only get a partial ordering of the monomials. A small enough generic perturbation of $\boldsymbol{\omega}$ refine the partial order to a term order.

We will now investigate how (regular) subdivisions can help to find generating sets and Gröbner bases of toric ideals. For a vector $\boldsymbol{v} \in \mathbb{Z}_{\geq 0}^{\mathcal{A}}$ we call the set $\operatorname{supp} \boldsymbol{v} := \{\boldsymbol{a} \in \mathcal{A} \mid v_{\boldsymbol{a}} \neq 0\}$ the *support* of $\boldsymbol{v}$ and say that $\boldsymbol{x}^{\boldsymbol{v}}$ is supported on $F \subseteq P$ whenever $\operatorname{supp} \boldsymbol{v} \subseteq F$. As a preliminary step, suppose P has a covering $\mathcal{C}$ by integrally closed polytopes (cf. Section 1.2.2). Then we can restrict the generating set (Section 2.2) of the vector space $\mathcal{I}_P$ to those binomials $\boldsymbol{x}^{\boldsymbol{u}} - \boldsymbol{x}^{\boldsymbol{v}}$ which have at least one monomial supported in a cell of $\mathcal{C}$.

For a subdivision $\mathcal{S}$ of $P \subset \mathbb{R}^d$ into lattice polytopes, we call $N \subseteq \mathcal{A}$ a non-face if $N \not\subseteq Q$ for all $Q \in \mathcal{S}$. (Our two notions of a non-face — for a triangulation and

[1] To be on the safe side, assume that the numbers $\omega_{\boldsymbol{a}}$ are linearly independent over $\mathbb{Q}$.

for a subdivision — agree for full triangulations.) If, for example, $\mathcal{S}$ comes from a lattice dicing as in Section 2.1.3, then all minimal non-faces have size two.

Given $\boldsymbol{\omega} \in \mathbb{R}^{\mathcal{A}}$ with induced regular subdivision $\mathcal{S}_{\boldsymbol{\omega}}$, and given a monomial $\boldsymbol{x}^{\boldsymbol{u}} \in S$, we call a monomial $\boldsymbol{x}^{\boldsymbol{v}}$ *standard*, written $\boldsymbol{x}^{\boldsymbol{v}} \in \operatorname{std}_{\boldsymbol{\omega}}(\boldsymbol{x}^{\boldsymbol{u}})$, if $\boldsymbol{x}^{\boldsymbol{u}} - \boldsymbol{x}^{\boldsymbol{v}} \in \mathfrak{I}_P$, and $\boldsymbol{v}$ minimizes $\langle \boldsymbol{\omega}, \boldsymbol{v} \rangle$ subject to this condition. Again, we can restrict the generating set (Section 2.2) of the vector space $\mathfrak{I}_P$ to binomials $\boldsymbol{x}^{\boldsymbol{u}} - \boldsymbol{x}^{\boldsymbol{v}}$ for which $\boldsymbol{x}^{\boldsymbol{v}} \in \operatorname{std}_{\boldsymbol{\omega}}(\boldsymbol{x}^{\boldsymbol{u}})$. If the cells of $\mathcal{S}_{\boldsymbol{\omega}}$ are integrally closed then the definition of regular subdivision says that $\boldsymbol{x}^{\boldsymbol{v}}$ is standard if and only if $\boldsymbol{x}^{\boldsymbol{v}}$ is supported on a cell of $\mathcal{S}_{\boldsymbol{\omega}}$. In particular, for every non-face $N \subseteq \mathcal{A}$ there is a monomial $\boldsymbol{x}^{\boldsymbol{v}}$ supported on a cell of $\mathcal{S}_{\boldsymbol{\omega}}$ with $f_N := \boldsymbol{x}^N - \boldsymbol{x}^{\boldsymbol{v}} \in \mathfrak{I}_P$, where $\boldsymbol{x}^N$ denotes the squarefree monomial $\prod_{\boldsymbol{a} \in \mathcal{A}} x_{\boldsymbol{a}}$ with support N.

For the following lemma, we consider the polynomial rings $\Bbbk[x_{\boldsymbol{a}} : \boldsymbol{a} \in \mathcal{A} \cap F]$ and their ideals $\mathfrak{I}_F$ as subsets of S.

A lifting function $\boldsymbol{\omega}$ producing the regular subdivision $\mathcal{S}_{\boldsymbol{\omega}}$ of $\mathcal{A}$ is *tight* if $(\boldsymbol{a}, \omega_{\boldsymbol{a}})$ lies in the boundary of $\tilde{P} := \operatorname{conv}(\boldsymbol{a} \times [\omega_{\boldsymbol{a}}, \infty) \ : \ \boldsymbol{a} \in \mathcal{A})$ for every $\boldsymbol{a} \in \mathcal{A}$. Observe that if $\omega_{\boldsymbol{a}}$ is not tight then there is a canonical way of making it tight: decrease the entries of $\omega_{\boldsymbol{a}}$ that are not tight until they are (without changing $\tilde{P}$). If all $\boldsymbol{a} \in \mathcal{A}$ are vertices of $\mathcal{S}_{\boldsymbol{\omega}}$, then $(\mathcal{S}_{\boldsymbol{\omega}}, \boldsymbol{\omega})$ is automatically tight.

LEMMA 2.20. *Suppose $\mathcal{S}_{\boldsymbol{\omega}}$ is a regular subdivision of the lattice polytope P into integrally closed lattice polytopes.*

(1) *The toric ideal $\mathfrak{I}_P$ is generated by $\mathfrak{I}_F$ for $F \in \mathcal{S}_{\boldsymbol{\omega}}$ together with f_N for minimal non-faces N.*
(2) *If $\boldsymbol{\omega}$ is tight for $\mathcal{S}_{\boldsymbol{\omega}}$ then for any small enough generic perturbation $\boldsymbol{\omega}'$ of $\boldsymbol{\omega}$, combining Gröbner bases for the $\mathfrak{I}_F$ with respect to $\boldsymbol{\omega}'|_F$ with the f_N for minimal non-faces N yields an $\boldsymbol{\omega}'$-Gröbner basis of $\mathfrak{I}_P$.*

Part (1) of this lemma will be used in Corollary 3.8 to show type **B** polytopes are quadratically generated. It follows readily from part (2).

PROOF. Take a binomial $f = \boldsymbol{x}^{\boldsymbol{u}} - \boldsymbol{x}^{\boldsymbol{v}} \in \mathfrak{I}_P$ with $\boldsymbol{x}^{\boldsymbol{u}} = \operatorname{in}_{\boldsymbol{\omega}'} f$.

If $\operatorname{supp} \boldsymbol{u}$ is a non-face of $\mathcal{S}_{\boldsymbol{\omega}}$, then there is a minimal non-face $N \subseteq \operatorname{supp} \boldsymbol{u}$, and $\operatorname{in}_{\boldsymbol{\omega}'} f_N = \boldsymbol{x}^N | \boldsymbol{x}^{\boldsymbol{u}} = \operatorname{in}_{\boldsymbol{\omega}'} f$.

If $\operatorname{supp} \boldsymbol{u}$ is contained in a face F of $\mathcal{S}_{\boldsymbol{\omega}}$, we claim that $\operatorname{supp} \boldsymbol{v}$ must also be contained in F: we have $\langle \boldsymbol{\omega}, \boldsymbol{u} \rangle \geq \langle \boldsymbol{\omega}, \boldsymbol{v} \rangle$ (as $\boldsymbol{\omega}'$ is a small perturbation of $\boldsymbol{\omega}$ and we have strict inequality for $\boldsymbol{\omega}'$), and $\sum_{\boldsymbol{a} \in \mathcal{A}} u_{\boldsymbol{a}} \boldsymbol{a} = \sum_{\boldsymbol{a} \in \mathcal{A}} v_{\boldsymbol{a}} \boldsymbol{a}$ is an affine dependence. So the tightness condition yields $\omega_{\boldsymbol{a}} = \langle \boldsymbol{\eta}_F, \boldsymbol{a} \rangle + \zeta_F$ for all $\boldsymbol{a} \in \operatorname{supp} \boldsymbol{v}$. □

If we apply the preceding lemma to a regular unimodular triangulation, we obtain the following corollary which is contained in [**132**, Corollaries 8.4, 8.8].

COROLLARY 2.21. *If $\mathcal{T}_{\boldsymbol{\omega}}$ is a regular unimodular triangulation $\mathcal{T}_{\boldsymbol{\omega}}$ of P, then*

$$\operatorname{in}_{\boldsymbol{\omega}} \mathfrak{I}_P = \left\langle \prod_{\boldsymbol{a} \in N} x_{\boldsymbol{a}} \ : N \textit{ is a minimal non-face of } \mathcal{T}_{\boldsymbol{\omega}} \right\rangle.$$

This ideal is known as the Stanley-Reisner ideal of the simplicial complex $\mathcal{T}_{\boldsymbol{\omega}}$. The formula allows us to recover $\mathcal{T}_{\boldsymbol{\omega}}$ from $\operatorname{in}_{\boldsymbol{\omega}} \mathfrak{I}_P$ (its faces correspond to the monomials not in $\operatorname{in}_{\boldsymbol{\omega}} \mathfrak{I}_P$, and vice versa). In fact, by Lemma 2.20 even a Gröbner basis can be read off of $\mathcal{T}_{\boldsymbol{\omega}}$.

If $\mathcal{T}_\omega$ is not unimodular then the following modified formula is still true:

$$\operatorname{Rad}(\operatorname{in}_\omega \mathcal{I}_P) = \left\langle \prod_{\boldsymbol{a} \in N} \boldsymbol{x_a} \ : N \text{ is a minimal non-face of } \mathcal{T}_\omega \right\rangle.$$

In particular, we can still recover $\mathcal{T}_\omega$ from $\operatorname{in}_\omega \mathcal{I}_P$, but not the other way around.

THEOREM 2.22. *Given that $\mathcal{A}$ generates the lattice $\mathbb{Z}^{d+1}$, the initial ideal $\operatorname{in}_\omega \mathcal{I}_P$ is squarefree if and only if the regular triangulation $\mathcal{T}_\omega$ of P is unimodular.*

This theorem is Corollary 8.9 in [**132**]; it follows from [**72**, Thm. 5.3], and is one of the primary motivations for studying regular unimodular triangulations, from the perspective of algebraic geometry.

PROOF. If $\mathcal{T}_\omega$ is unimodular, the previous lemma shows that $\operatorname{in}_\omega \mathcal{I}_P$ is squarefree.

So, assume $\mathcal{T}_\omega$ is not unimodular, and let $F = \operatorname{conv}(\boldsymbol{a}_0, \dots, \boldsymbol{a}_d) \in \mathcal{T}_\omega$ be a simplex of determinant $D > 1$. Let Λ denote the (strict) sublattice of $\mathbb{Z}^{d+1}$ generated by the vertices of F. Observe that $D\boldsymbol{m} \in \Lambda$ for all $\boldsymbol{m} \in \mathbb{Z}^{d+1}$.

We will construct a vector $\boldsymbol{b} \in \operatorname{cone} F \cap \mathbb{Z}^{d+1} \setminus \Lambda$ which is a non-negative integral linear combination of $\mathcal{A}$. First, choose $\boldsymbol{b}' \in \mathbb{Z}^{d+1} \setminus \Lambda$. By assumption, $\boldsymbol{b}'$ is an integral linear combination of $\mathcal{A}$. Adding a sufficiently large multiple of $D\sum_{\boldsymbol{a}\in\mathcal{A}} \boldsymbol{a}$ will make the coefficients non-negative. Then adding a sufficiently multiple of $\sum_{\boldsymbol{a}\in F} \boldsymbol{a}$ will yield a point in cone F.

Among all $\boldsymbol{n} \in \mathbb{Z}^{\mathcal{A}}_{\geq 0}$ satisfying $\sum_{\boldsymbol{a}\in\mathcal{A}} n_{\boldsymbol{a}} \boldsymbol{a} = \boldsymbol{b}$ choose the one with minimal $\boldsymbol{\omega}$-weight. Since $\boldsymbol{b} \notin \Lambda$, $\boldsymbol{x}^{\boldsymbol{n}}$ is not supported on F. Still, $\boldsymbol{x}^{\boldsymbol{n}}$ is never a leading term: $\boldsymbol{x}^{\boldsymbol{n}} \notin \operatorname{in}_\omega \mathcal{I}_P$.

Yet, $D\boldsymbol{b} \in \operatorname{cone} F \cap \Lambda$, so $D\boldsymbol{b} = \sum_{i=0}^{d} m_i \boldsymbol{a}_i$ for some $\boldsymbol{m}$, and $\boldsymbol{x}^{D\boldsymbol{n}} - \boldsymbol{x}^{\boldsymbol{m}} \in \mathcal{I}_P$. As $\boldsymbol{x}^{\boldsymbol{m}}$ is supported on the face F, it cannot be the leading term and $(\boldsymbol{x}^{\boldsymbol{n}})^D \in \operatorname{in}_\omega \mathcal{I}_P$. So $\operatorname{in}_\omega \mathcal{I}_P$ is not squarefree. □

Theorem 2.22 provides a method for constructing regular unimodular triangulations. Conversely, all regular unimodular triangulations constructed in the present article yield Gröbner bases of the corresponding toric ideals. Both directions of the theorem have been used – compare sections 3.3.1 and 3.2.3.

EXAMPLE 2.23. Consider the the twisted cubic curve. Let $P = [1,4]$ be a 1-dimensional polytope whose lattice point set and toric ideal are $\mathcal{A} = \{(1,1), (2,1), (3,1), (4,1)\}$ and $\mathcal{I}_P = \langle x_1x_3 - x_2^2, x_2x_4 - x_3^2, x_1x_4 - x_2x_3\rangle$. The different choices of $\boldsymbol{\omega} = (\omega_1, \omega_2, \omega_3, \omega_4)$ produce eight monomial initial ideals and four triangulations, depending on the values of $\lambda_1 := \omega_1 - 2\omega_2 + \omega_3$ and $\lambda_4 := \omega_4 - 2\omega_3 + \omega_2$:

$\boldsymbol{\omega}$			$\operatorname{in}_\omega \mathcal{I}_P$	$\mathcal{T}_\omega$
$\lambda_1 > 0$	and	$\lambda_4 > 0$	$\langle x_1x_3, x_2x_4, x_1x_4\rangle$	$[1,2], [2,3], [3,4]$
$\lambda_1 < 0$	and	$2\lambda_1 + \lambda_4 > 0$	$\langle x_1x_4, x_2^2, x_2x_4\rangle$	$[1,3], [3,4]$
$\lambda_1 + 2\lambda_4 > 0$	and	$2\lambda_1 + \lambda_4 < 0$	$\langle x_2^2, x_2x_3, x_2x_4, x_1x_4^2\rangle$	$[1,3], [3,4]$
$\lambda_1 + 2\lambda_4 < 0$	and	$\lambda_4 > 0$	$\langle x_2^2, x_2x_3, x_2x_4, x_3^3\rangle$	$[1,4]$
$\lambda_1 < 0$	and	$\lambda_4 < 0$	$\langle x_2^2, x_2x_3, x_3^2\rangle$	$[1,4]$
$\lambda_1 > 0$	and	$2\lambda_1 + \lambda_4 < 0$	$\langle x_1x_3, x_2x_3, x_2^3, x_3^2\rangle$	$[1,4]$
$\lambda_1 + 2\lambda_4 < 0$	and	$2\lambda_1 + \lambda_4 > 0$	$\langle x_1x_3, x_2x_3, x_1^2x_4, x_3^2\rangle$	$[1,2], [2,4]$
$\lambda_1 + 2\lambda_4 > 0$	and	$\lambda_4 < 0$	$\langle x_1x_3, x_1x_4, x_3^2\rangle$	$[1,2], [2,4]$

Observe how each triangulation corresponds to the (radical of the) initial ideal. The converse works precisely in the first case where the initial ideal is squarefree and the triangulation is unimodular.

2.4.2. Quadratic triangulations. If a polytope has a regular unimodular triangulation, we have seen that the size of the minimal non-faces controls the degree of the corresponding Gröbner basis. Of particular interest is the case of degree two – quadratic triangulations – especially given the connection to Koszul algebras. Recall that an algebra R over a field $\Bbbk$ is Koszul if $\Bbbk$ has a linear free resolution as an R–module. In this context, the second hierarchy of properties in section 1.2.5 is expressed in the following proposition (see [**28**, Cor. 2.1.3]).

PROPOSITION 2.24. *$\mathcal{I}_A$ has a quadratic initial ideal $\Rightarrow$ $\boldsymbol{k}[\boldsymbol{x}]/\mathcal{I}_A$ is Koszul $\Rightarrow$ $\mathcal{I}_A$ is generated by quadratic binomials.*

For polygons we have the following nice characterization by Bruns, Gubeladze, and Trung.

PROPOSITION 2.25 ([**28**, Cor. 3.2.5]). *A lattice polygon with at least four boundary lattice points has a quadratic triangulation.*

A non-unimodular polygon with exactly three boundary lattice points cannot have a quadratic triangulation because one cannot get rid of the cubic generator of $\mathcal{I}_A$ coming from the product of the corresponding three variables.

Lemma 2.21 also implies the following correspondence between quadratic triangulations and quadratic Gröbner bases referenced in Section 1.2.3.

THEOREM 2.26. *If P has a quadratic triangulation $\mathcal{T}$, then the defining ideal $\mathcal{I}_P$ of the projective toric variety $X_P \subset \mathbb{P}^{r-1}$ has a quadratic Gröbner basis. In this case, $\operatorname{in}(\mathcal{I}_P) = \langle x_{\boldsymbol{a}} x_{\boldsymbol{b}} \mid \boldsymbol{ab}$ is not an edge in $\mathcal{T}\rangle$. In particular, R_P is Koszul.*

See [**29**] for a collection of unsolved problems in the field.

CHAPTER 3

Examples

Here we present what is known (and unknown) for some particular families of polytopes. Most of them are connected to one of the classical crystallographic root systems. We will examine two distinct ways of associating polytopes to a root system Γ. In Section 3.1 we consider polytopes with facet normals in Γ (*polytopes of type* Γ), and in Section 3.2 polytopes with vertices in Γ (Γ-*root polytopes*). Then, in Section 3.3 we look at polytopes defined from graphs, including flow polytopes of directed graphs and characteristic polytopes of undirected ones. Finally, Section 3.5 examines smooth polytopes.

A (real, finite) *root system* is a family of vectors Γ that is invariant under reflection with respect to the hyperplanes orthogonal to each of the elements in Γ. It is *crystallographic* if Γ generates a lattice, which we denote Λ_Γ. Each system studied here satisfies that property, so when we refer to root systems, they are assumed to be crystallographic.

The direct sum

$$\Gamma \oplus \Gamma' := \Gamma \times \{0\} \cup \{0\} \times \Gamma'$$

of two root systems is a root system. Root systems that cannot be decomposed in this fashion are called *irreducible* and are classified as follows. (Here we are only considering crystallographic ones).

There are the four infinite families, that exist in all dimensions:

$$\begin{aligned}
\mathbf{A}_{n-1} &:= \{\boldsymbol{e}_i - \boldsymbol{e}_j \mid 1 \leq i, j \leq n\} \\
\mathbf{B}_n &:= \{\pm \boldsymbol{e}_i \mid 1 \leq i \leq n\} \cup \{\pm \boldsymbol{e}_i \pm \boldsymbol{e}_j \mid 1 \leq i < j \leq n\} \\
\mathbf{C}_n &:= \{\pm 2\boldsymbol{e}_i \mid 1 \leq i \leq n\} \cup \{\pm \boldsymbol{e}_i \pm \boldsymbol{e}_j \mid 1 \leq i < j \leq n\} \\
\mathbf{D}_n &:= \{\pm \boldsymbol{e}_i \pm \boldsymbol{e}_j \mid 1 \leq i < j \leq n\},
\end{aligned}$$

where, as usual, $\boldsymbol{e}_i$ is the i-th standard unit vector in $\mathbb{R}^n$.

In addition to these, there are five other root systems:

- The system $\mathbf{F}_4$ is spanned by the roots in $\mathbf{B}_4$ together with all vectors in $\{(\pm{}^1/_2, \pm{}^1/_2, \pm{}^1/_2, \pm{}^1/_2)\}$.
- The root system $\mathbf{E}_8$ is the union of $\mathbf{D}_8$ and

$$\{\boldsymbol{x} \mid x_i = \pm{}^1/_2 \text{ for } 1 \leq i \leq 8 \text{ and } \textstyle\sum x_i \text{ even}\}.$$

- The root system $\mathbf{E}_7$ is the subset of $\mathbf{E}_8$ of all vectors whose entries sum up to zero.
- The root system $\mathbf{E}_6$ is the set of roots of $\mathbf{E}_8$ that are spanned by the vector

$$({}^1/_2, \quad -{}^1/_2, \quad -{}^1/_2, \quad -{}^1/_2, \quad -{}^1/_2, \quad -{}^1/_2, \quad -{}^1/_2, \quad {}^1/_2)$$

together with $\boldsymbol{e}_1 + \boldsymbol{e}_2$ and $-\boldsymbol{e}_i + \boldsymbol{e}_{i+1}$ for $1 \le i \le 4$.

- Finally, the root system $\mathbf{G}_2$ is the set of vectors

$$\begin{array}{lll} \pm(1\ \ -1\ \ \ 0), & \pm(1\ \ \ 0\ \ -1), & \pm(0\ \ 1\ \ -1), \\ \pm(2\ \ -1\ \ -1), & \pm(1\ \ -2\ \ \ 1), & \pm(1\ \ 1\ \ -2). \end{array}$$

For each root system $\Gamma \in \{\mathbf{A}_{n-1}, \mathbf{B}_n, \mathbf{C}_n, \mathbf{D}_n, \mathbf{F}_4\}$ we define the special choice of *positive* roots in the root system as the subset $\Gamma^+ \subset \Gamma$ consisting of all roots whose first nonzero entry is positive, and $\tilde{\Gamma}$ and $\tilde{\Gamma}^+$ are defined as $\Gamma \cup \{\mathbf{0}\}$ and $\Gamma^+ \cup \{\mathbf{0}\}$, respectively.

3.1. Polytopes cut out by roots

Given a root system Γ, root hyperplanes are hyperplanes of the form $H_{\boldsymbol{v},b} := \{\boldsymbol{v}\boldsymbol{x} \le w\}$ where $\boldsymbol{v} \in \Gamma$ and $w \in \mathbb{Z}$. If Γ is a crystallographic root system with lattice Λ_Γ then the root hyperplanes are lattice hyperplanes for the dual lattice

$$\Lambda_\Gamma^* := \{\boldsymbol{x} : \boldsymbol{v}\boldsymbol{x} \in \mathbb{Z}, \forall \boldsymbol{v} \in \Gamma\}.$$

A polytope is *of type* Γ if it is a lattice polytope (for the lattice Λ_Γ^*) and all its facet-defining hyperplanes are root hyperplanes. Note that polytopes cut out by a non-irreducible root system are products of their irreducible components.

The following is a general result about polytopes cut out by roots (cf. [**66**, p. 90]).

LEMMA 3.1. *Given an irreducible root system Γ spanning $\mathbb{R}^d$, every cell in the hyperplane arrangement given by the (infinite) family of hyperplanes of the form $\{\boldsymbol{v}x = z\}$ for $\boldsymbol{v} \in \Gamma$ and $z \in \mathbb{Z}$ is a simplex.*

That is, the arrangement of all root hyperplanes is an (infinite, periodic) triangulation of $\mathbb{R}^d$. Its cells are called *alcoves*. It may, however, have vertices which do not belong to Λ_Γ^*. For example, for the root system $\mathbf{B}_2$, $\Lambda = \Lambda^* = \mathbb{Z}^2$, but the arrangement of root hyperplanes contains vertices in $(1/2, 1/2) + \mathbb{Z}^2$.

Payne proved another important general result about polytopes cut out by roots using Frobenius splittings:

THEOREM 3.2. [**106**, Thm. 1.1] *Every type Γ polytope for each of the classical root systems $\mathbf{A}_{n-1}$, $\mathbf{B}_n$, $\mathbf{C}_n$ and $\mathbf{D}_n$ is integrally closed and Koszul.*

3.1.1. Type A Polytopes. The type **A** root system is very special in that the matrix whose columns are its roots is totally unimodular. In particular, all polytopes of type **A** are totally unimodular. Therefore, they have quadratic triangulations which can be realized via the construction in Theorem 2.4. The following statement (without the proof of flagness) appeared already in [**77**, Lemma 2.4].

THEOREM 3.3. *Let P be a polytope of type* **A**. *The lattice dicing subdivision $\mathcal{T}$ obtained from slicing P by all the lattice hyperplanes with normal in $\mathbf{A}_n$ is a quadratic triangulation of P.*

PROOF. Theorem 2.4 established that $\mathcal{T}$ is a regular lattice subdivision in which all faces are compressed, and Lemma 3.1 shows that this is in fact a triangulation. So, it only remains to be shown that it is flag.

For this, suppose N is a minimal non-face with more than two elements. Since N is not a face, the relative interior of $\operatorname{conv}(N)$ is cut by some hyperplane H spanned by a face of $\mathcal{T}$, which must be a root hyperplane. In particular, N contains points

$\boldsymbol{n}^+$ and $\boldsymbol{n}^-$ on both sides of that hyperplane, which contradicts the fact that $\boldsymbol{n}^+\boldsymbol{n}^-$ is an edge. □

We now discuss two particularly interesting cases of type **A** polytopes, order polytopes and hypersimplices.

Let $(X, \preccurlyeq)$ be a partial order on $X = \{1, \ldots, n\}$. A vector $\boldsymbol{v} \in [0,1]^n$ is said to *respect the order* if $v_i \leq v_j$ whenever $i \preccurlyeq j$. A *linear extension* of $\preccurlyeq$ is a total order on X that refines the partial order. The *order polytope* associated $(X, \preccurlyeq)$ is the polytope

$$O(\preccurlyeq) := \{\boldsymbol{x} \in [0,1]^n \mid \boldsymbol{x} \text{ respects the order}\}.$$

Vertices of $O(\preccurlyeq)$ are the characteristic vectors of *up-close subsets* or *filters* of the poset (that is, sets $S \subset X$ with the property that $i \in S$ and $i \preccurlyeq j$ implies $j \in S$). Facets of $O(\preccurlyeq)$ are in bijection with *covering relations*, *minimal elements* and *maximal elements.* If $i \preccurlyeq j$ is a covering relation (meaning that there is no k with $i \preccurlyeq k \preccurlyeq j$) then $x_i \leq x_j$ defines a facet, and if i is a minimal (resp. maximal) element then $x_i \geq 0$ (resp. $x_i \leq 1$) defines a facet. All facets are of one of these forms. In particular, the facet vectors of an order polytope are contained in the set of vectors $\{e_i : i \in [n]\} \cup \{e_i - e_j : i, j \in [n]\}$, which is mapped to $\mathbf{A}_n$ by the linear isomorphism $e_i \mapsto e_i - e_{n+1}$. This shows that $O(\preccurlyeq)$ is of type **A**, and also that it is compressed.

The quadratic triangulation of an order polytope guaranteed by Theorem 3.3 was first studied by Stanley [**127**]. Its maximal simplices are in bijection to the linear extensions of the partial order. In particular, the normalized volume of $O(\preccurlyeq)$ equals the number of distinct linear extensions of $\preccurlyeq$.

The d-dimensional hypersimplex $\Delta(d, k)$ for $1 \leq k \leq d$ is defined as

$$\Delta(d,k) := \{\boldsymbol{x} \in \mathbb{R}^d \mid k - 1 \leq \sum_i x_i \leq k, \, \mathbf{0} \leq \boldsymbol{x} \leq \mathbf{1}\}.$$

Alternatively, each $\Delta(d, k)$ can also be realized as the intersection of the $(d+1)$ dimensional 0/1-cube with the hyperplane $\{\boldsymbol{x} \mid \sum_i x_i = k\}$. The hypersimplices for $k = 1$ and $k = d$ are simplices. The hypersimplex $\Delta(3, 2)$ is the octahedron. To see that these polytopes are root system polytopes of type $\mathbf{A}_{d-1}$, apply the unimodular transformation given by $y_j := \sum_{i=1}^{j} x_i$. The facet inequalities are then given by $y_j - y_{j-1} \geq 0$ and $k - 1 \leq y_d \leq k$. The fact that these hypersimplices are also compressed follows directly from condition (3) of Theorem 2.3.

3.1.2. Type B Polytopes. The root system $\mathbf{B}_n \subset \mathbb{R}^n$ has two types of roots. Their distinction appears in our proofs of Proposition 3.4 and Lemma 3.5. The "short roots" are given by the vectors $\pm\boldsymbol{e}_i$, and the "long roots" are given by the vectors $\pm\boldsymbol{e}_i \pm \boldsymbol{e}_j$ for $1 \leq i, j \leq n$.

PROPOSITION 3.4. *If P is a lattice polytope with facet normals in $\mathbf{B}_n$, then P admits a regular unimodular triangulation.*

The proof requires the following lemma.

LEMMA 3.5. *Let M be a matrix with rows in $\mathbf{B}_n$, N be a matrix with rows in $\{e_i : i \in [n]\}$ (the set of short roots), and $\boldsymbol{w} \in \mathbb{Z}^k$, where k is the number of rows in N. If the system*

$$M\boldsymbol{x} = 0, \qquad N\boldsymbol{x} = \boldsymbol{w}$$

has a unique solution, then this solution is integral.

PROOF. First, observe that there is no loss of generality in assuming that all rows of M are long roots (as all rows that are short roots can be put in N). Also, if a long root $\boldsymbol{e}_i \pm \boldsymbol{e}_j$ in M shares a coordinate with a short root $\boldsymbol{e}_i$ in N, then the other coordinate of that long root is fixed to the value $x_j = \pm w_i$, so removing that row from M and putting a new row in N, yields equivalent system.

After this process has been applied as many times as possible, M and N operate in disjoint sets of coordinates. At which point, setting all coordinates in N to their value w_i and all coordinates not in N to zero gives an integral solution of the system (unless N is inconsistent, in which case the original system had no solution). □

PROOF OF PROPOSITION 3.4. Slice P by all short roots. This gives a regular subdivision of P. The proposition follows from showing that the cells in this subdivision have integral vertices and are compressed.

Let $\boldsymbol{a}$ be a vertex of this subdivision and let $\boldsymbol{b}$ be a vertex of the carrier face of $\boldsymbol{a}$ in P. Then, Lemma 3.5 implies that $\boldsymbol{a} - \boldsymbol{b}$ is integral, since it is determined by setting some coordinates to a fixed value and homogenized versions of some of the facet-defining inequalities of P. Thus $\boldsymbol{a} = \boldsymbol{b} + (\boldsymbol{a} - \boldsymbol{b})$ is also integral.

It remains to show that all cells in the subdivision are compressed. By construction, the cells have width one in the direction of the short roots. In the direction of a long root $\pm\boldsymbol{e}_i \pm \boldsymbol{e}_j$ consider the projection onto the x_i-x_j-plane. In each case the cell has width one. □

EXAMPLE 3.6. The quadratic triangulations obtained for type **A** polytopes in Theorem 3.3 are in fact "type **A** triangulations", in the sense that they consist of type **A** simplices. The same does not in general occur in type **B**.

For example, consider the following type **B** polytope:

$$P = \operatorname{conv}\begin{bmatrix} 0&1&0&1&1&1\\ 0&0&1&1&1&0\\ 0&0&0&0&1&-1 \end{bmatrix} = \left\{\boldsymbol{x} \in [0,1]^3 \ : \ z \le x, y; x + z \ge 0; y - z \ge 1\right\}$$

The only type **B** hyperplane cutting through its interior but not creating any non-integer vertices is $z = 0$. This is the hyperplane used in the proof of Proposition 3.4, and cuts P into two compressed type **B** polytopes. Each of them is a square pyramid and they both have type **B** unimodular triangulations, but their type **B** triangulations use different diagonals of the square. Meaning, in the last step of the proof we cannot triangulate the cells of the short-root dicing of P using transformed **A**-triangulations.

We do not know whether Proposition 3.4 can be extended to yield quadratic triangulations of type **B** polytopes. We do know arbitrary pulling refinements will not work, because there are non-flag pulling triangulations of the order polytope $[0,1]^3$. Even pulling according to the vertex order of a linear functional may produce non-flag triangulations in higher dimensional order polytopes. (The smallest example we know of is six-dimensional: the order polytope of the Boolean lattice on three elements with the minimum and maximum elements removed.)

However, the following lemma, combined with Lemma 2.20 does guarantee that the toric ideals of type **B** polytopes are quadratically generated, since **B**-cells are equivalent to **A**-cells (which have quadratic triangulations).

LEMMA 3.7. *If P is a lattice polytope with facet normals in $\mathbf{B}_n$ and Q is a cell in the lattice dicing subdivision of P obtained from slicing P by all the lattice hyperplanes with normals among the short roots in $\mathbf{B}_n$, then Q is equivalent to an order polytope.*

PROOF. We can assume Q is full-dimensional. Up to translation, Q is given by the intersection of $[0,1]^n$ with some constraints $x_i \lessgtr x_j$ and some $x_i + x_j \lessgtr 1$. In either case, the cell must contain the point $\boldsymbol{b} := \frac{1}{2}\mathbf{1}$. Since $\boldsymbol{b}$ cannot be a vertex, there must be a vertex $\boldsymbol{a} \in \{0,1\}^d$ such that for small $\epsilon > 0$, the points $\boldsymbol{b} \pm \epsilon(\boldsymbol{a} - \boldsymbol{b})$ both belong to our cell. Due to the nature of its inequalities, our cell contains the long diagonal from $\boldsymbol{a}$ to $\mathbf{1} - \boldsymbol{a}$. So, mapping x_i to $1 - x_i$ whenever $a_i = 1$ identifies the cell unimodularly with an order polytope. □

COROLLARY 3.8. *Toric ideals of type* **B** *polytopes are quadratically generated.*

3.1.3. Type $\mathbf{C}_n$ and $\mathbf{D}_n$ Polytopes. Payne's theorem (Theorem 3.2) guarantees that all polytopes of type **C** and **D** are integrally closed. However, we do not know whether they all have unimodular triangulations for cases other than $\mathbf{D}_n \cong \mathbf{A}_n$ $(n \leq 3)$ and $\mathbf{C}_n \cong \mathbf{B}_n$ $(n \leq 2)$.

3.1.4. Type F Polytopes. Polytopes with facets in $\mathbf{F}_4$ do not in general have regular unimodular triangulations. For an example, consider the polytope $P_{\mathbf{F}_4}$ defined by the following linear inequalities:

$$
\begin{array}{lll}
x_4 + x_2 \leq 0 & x_4 - x_1 \geq 0 & x_4 - x_2 \leq 0 \\
x_3 + x_1 \geq 0 & x_3 - x_1 \leq 2 &
\end{array}
$$

The dual lattice $(\mathbf{F}_4)^*$ is the sublattice of $\mathbb{Z}^4$ containing all points with even coordinate sum. Hence, the lattice points in P are

$$
\begin{array}{ll}
(-1,\ 0,\ 1,\ 0) & (-1,-1,\ 1,-1) \\
(\ 0,\ 0,\ 0,\ 0) & (-1,\ 1,\ 1,-1) \\
(\ 0,\ 0,\ 2,\ 0) &
\end{array}
$$

which are also the vertices of $P_{\mathbf{F}_4}$. In particular, since $P_{\mathbf{F}_4}$ is a non-unimodular empty simplex, it does not have a unimodular triangulation. In fact, it is not even integrally closed. For example, $(-1,0,2,-1) \in (\mathbf{F}_4)^* \cap 2P$ is not a sum of two lattice points in $P_{(\mathbf{F}_4,\mathbf{c})}$. Therefore, polytopes cut out by $\mathbf{F}_4$ are not in general integrally closed, which answers a question left open in [**106**].

3.1.5. Type $\mathbf{E}_6$, $\mathbf{E}_7$, $\mathbf{E}_8$ Polytopes. Here is an example of a type $\mathbf{E}_8$ polytope that does not have a regular unimodular triangulation. $P_{\mathbf{E}_8}$ is the product of the polytope $P_{\mathbf{F}_4}$ defined in Section 3.1.4 with itself. The facet normals of this polytope are roots in $\mathbf{E}_8$. The 25 vertices of $P_{\mathbf{E}_8}$ are the products of the five vertices

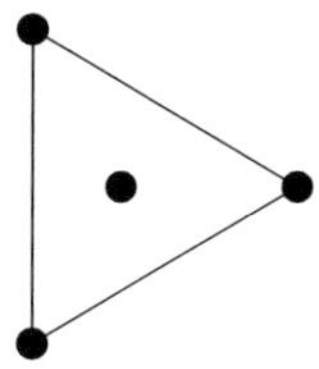

FIGURE 3.1. A triangle without a quadratic triangulation

of $P_{\mathbf{F}_4}$. Additionally, we have the following 12 lattice points:

$$\begin{array}{rrrrrrrr}
(-1 & 0 & 1 & -1 & -1 & 0 & 1 & -1) \\
(-1 & 0 & 1 & -1 & 0 & 0 & 1 & 0) \\
(\ 0 & 0 & 1 & 0 & -1 & 0 & 1 & -1) \\
(\ 0 & 0 & 1 & 0 & 0 & 0 & 1 & 0) \\
(-1/2 & -1/2 & 1/2 & -1/2 & -1/2 & -1/2 & 1/2 & -1/2) \\
(-1/2 & -1/2 & 1/2 & -1/2 & -1/2 & 1/2 & 3/2 & -1/2) \\
(-1/2 & 1/2 & 1/2 & -1/2 & -1/2 & -1/2 & 3/2 & -1/2) \\
(-1/2 & 1/2 & 1/2 & -1/2 & -1/2 & 1/2 & 1/2 & -1/2) \\
(-1/2 & -1/2 & 3/2 & -1/2 & -1/2 & -1/2 & 3/2 & -1/2) \\
(-1/2 & -1/2 & 3/2 & -1/2 & -1/2 & 1/2 & 1/2 & -1/2) \\
(-1/2 & 1/2 & 3/2 & -1/2 & -1/2 & -1/2 & 1/2 & -1/2) \\
(-1/2 & 1/2 & 3/2 & -1/2 & -1/2 & 1/2 & 3/2 & -1/2)
\end{array}$$

The point $(-2,-1,2,-1,-1,0,2,-1)$ is in twice the polytope $P_{\mathbf{E}_8}$, but it is not the sum of two lattice points in $P_{\mathbf{E}_8}$. So $P_{\mathbf{E}_8}$ is not integrally closed, and cannot have a regular unimodular triangulation.

The $\mathbf{E}_7$ and $\mathbf{E}_6$ cases remain open.

3.1.6. Type G Polytopes. Every type $\mathbf{G}_2$ polytope $P_{\mathbf{G}_2}$ is a polygon. As such, they have regular unimodular triangulations. By Proposition 2.25 they even have quadratic triangulations whenever there are at least four boundary lattice points. Up to lattice equivalence, the only type $\mathbf{G}_2$ polygons with three boundary lattice points are a unimodular triangle and the triangle of Figure 3.1.

3.2. Polytopes spanned by roots

Now we consider polytopes defined as the convex hull of a subset of one of the root systems. Polytopes of this type are widely studied, see e.g. Gelfand, Graev, and Postnikov [**55**], Ohsugi and Hibi [**100, 101**], Meszaros [**86, 87**], and Cho [**37**]. Ohsugi and Hibi have found regular unimodular triangulations for several classes of these polytopes.

Before presenting results, we introduce some notation. Let $\mathcal{A}$ be a set of vectors in $\mathbb{R}^n$ and $\widetilde{\mathcal{A}} := \mathcal{A} \cup \{\mathbf{0}\}$. The polytope associated to this configuration is $P_{\mathcal{A}} := \operatorname{conv}(\mathcal{A})$.

3.2.1. Sub-configurations of $\mathbf{A}_n$: arc polytopes. Arc polytopes are the easiest type of a root polytope to find unimodular triangulations for. Since $\mathbf{A}_n$ is totally unimodular as a vector configuration, for every $\mathcal{A} \subseteq \mathbf{A}_n$ each simplex containing the origin as a vertex is unimodular. As a result:

THEOREM 3.9. *The following polytopes have regular unimodular triangulations.*

(1) $P_{\widetilde{\mathcal{A}}}$ *for any* $\mathcal{A} \subseteq \mathbf{A}_{n-1}$. *In particular,* $P_{\mathcal{A}}$ *for any* $\mathcal{A} \subseteq \mathbf{A}_{n-1}$ *with* $\mathbf{0} \in P_{\mathcal{A}}$.
(2) $P_{\mathcal{A}}$ *for any* $\mathcal{A} \subseteq \mathbf{A}_{n-1}$ *with* $\mathbf{0} \notin \operatorname{aff}(\mathcal{A})$. *In this case* every *triangulation is unimodular.*

PROOF. Part (1) follows from Stanley [**126**, Ex. 2.4a]. For any pulling triangulation in which $\mathbf{0}$ is the first point pulled, each simplex in the triangulation will contain $\mathbf{0}$. By unimodularity of the vertex matrix, this means every resulting simplex has volume one.

The argument for Part (2) is essentially the same. $Q := \operatorname{conv}(\{\mathbf{0}\} \cup \mathcal{A}^+)$ is a pyramid over $P_{\mathcal{A}^+}$ with apex $\mathbf{0}$, so every simplex of every triangulation of it contains $\mathbf{0}$, and is hence unimodular. □

The only configurations $\mathcal{A} \subseteq \mathbf{A}_{n-1}$ not addressed by Theorem 3.9 are those that have $\mathbf{0}$ in their affine hull, but not in their convex hull. Not all of these have a regular unimodular triangulation. In fact, the polytope of such a configuration can even fail to be integrally closed.

EXAMPLE 3.10. Consider the subconfiguration of $\mathbf{A}_4^+$ consisting of $\boldsymbol{e}_1 - \boldsymbol{e}_2$, $\boldsymbol{e}_2 - \boldsymbol{e}_3$, $\boldsymbol{e}_3 - \boldsymbol{e}_4$, $\boldsymbol{e}_4 - \boldsymbol{e}_5$, $\boldsymbol{e}_1 - \boldsymbol{e}_4$ and $\boldsymbol{e}_3 - \boldsymbol{e}_5$. It is 4-dimensional with six vertices, so it has a unique affine dependence. Namely:

$$(\boldsymbol{e}_1 - \boldsymbol{e}_2) + (\boldsymbol{e}_2 - \boldsymbol{e}_3) + 2(\boldsymbol{e}_3 - \boldsymbol{e}_5) = (\boldsymbol{e}_1 - \boldsymbol{e}_4) + (\boldsymbol{e}_3 - \boldsymbol{e}_4) + 2(\boldsymbol{e}_4 - \boldsymbol{e}_5).$$

Since both sides of the dependence have some coefficient different from one, neither triangulation of this circuit is unimodular (part (1) of Lemma 2.6).

Even when there is a regular unimodular triangulation, it will not necessarily be flag.

EXAMPLE 3.11. Consider the subconfiguration of $\mathbf{A}_5^+$ consisting of $\boldsymbol{e}_1 - \boldsymbol{e}_2$, $\boldsymbol{e}_2 - \boldsymbol{e}_3$, $\boldsymbol{e}_1 - \boldsymbol{e}_3$, $\boldsymbol{e}_3 - \boldsymbol{e}_5$, $\boldsymbol{e}_3 - \boldsymbol{e}_4$, and $\boldsymbol{e}_4 - \boldsymbol{e}_5$. This is again a circuit and its only affine dependence is

$$(\boldsymbol{e}_1 - \boldsymbol{e}_2) + (\boldsymbol{e}_2 - \boldsymbol{e}_3) + (\boldsymbol{e}_3 - \boldsymbol{e}_5) = (\boldsymbol{e}_1 - \boldsymbol{e}_3) + (\boldsymbol{e}_3 - \boldsymbol{e}_4) + (\boldsymbol{e}_4 - \boldsymbol{e}_5).$$

Since the circuit has three points on each side, none of its triangulations is flag (part (2) of Lemma 2.6). Let us mention that this configuration is lattice equivalent to the set of vertices of the third Birkhoff polytope. (See Theorem 3.24 and the discussion before it.)

Notice that Theorem 3.9(1) implies $P_{\mathbf{A}_{n-1}}$ has a regular unimodular triangulation. Even more is known for the subconfiguration of all positive roots.

THEOREM 3.12. (1) $P_{\widetilde{\mathbf{A}}_{n-1}^+}$ *has a quadratic triangulation.*
(2) $P_{\mathbf{A}_{n-1}^+}$ *has a regular unimodular triangulation in which all non-faces have size two or three.*

Part (1) was proved by Gelfand, Graev, and Postnikov [**55**], and part (2) is due to Kitamura [**76**]. This statement is the best possible since $\mathbf{A}_{n-1}^+$ cannot have a quadratic triangulation for $n \geq 6$, as demonstrated in the following example.

EXAMPLE 3.13 (Example 3.11 continued). The vectors $\boldsymbol{e}_1 - \boldsymbol{e}_2$, $\boldsymbol{e}_1 - \boldsymbol{e}_3$, $\boldsymbol{e}_2 - \boldsymbol{e}_3$, $\boldsymbol{e}_4 - \boldsymbol{e}_5$, $\boldsymbol{e}_4 - \boldsymbol{e}_6$, $\boldsymbol{e}_5 - \boldsymbol{e}_6$ are the vertex set of the face of $\mathbf{A}_5^+$ defined by $x_1 + x_2 + x_3 = 0$. Any flag triangulation of $\mathbf{A}_5^+$ would in particular give a flag triangulation of that face, but that face does not have any flag triangulation.

A-root configurations have a natural graph-theoretic interpretation. Let $G = (V, A)$ be a directed graph, which we assume to be connected in the undirected sense. Its (directed) incidence matrix is the $(|V| \times |A|)$-matrix D_G with a 1 at position (v, a) if v is the head of the directed edge (arc) a, with a -1 if v is the tail of a, and with a 0 otherwise. The *arc polytope* of G is the convex hull of the columns of D_G. By construction, the columns are roots of type $\mathbf{A}_{n-1}$. Other than possibly the origin, the only lattice points in the arc polytope are its vertices, which correspond to arcs in G.

The following easy properties were noticed by Hibi and Ohsugi [**101**].

LEMMA 3.14. *Let G be a connected graph with n vertices, and $\mathcal{A}$ be the set of lattice points in its arc polytope.*

(1) *$\mathcal{A}$ contains the origin if and only if G has a directed cycle.*
(2) *$\mathcal{A}$ has dimension $d-1$ (that is, it affinely spans the hyperplane $\sum x_i = 0$) if and only if G has an* unbalanced cycle. *(Here, an undirected cycle in a directed graph is called* balanced *if it has the same number of edges oriented in both directions.)*
(3) *$\mathcal{A}$ is totally unimodular (all its full-dimensional simplices, and hence all its triangulations, are unimodular) if and only if all unbalanced cycles contain exactly one more edge oriented in one direction than in the other.* [**100**, Lemma 3.6]

In part (3) we mean totally unimodular with respect to the root lattice. This is different from the convention in [**101**] where unimodular is meant with respect to the lattice generated by $\mathcal{A}$.

In part (2) observe that balanced cycles must be even, so that the graphs having only balanced cycles must be bipartite. An important subclass are those where all edges are directed from one part to the other. In this case the arc polytopes are subpolytopes of the product of two simplices, so all their triangulations are unimodular. (This follows also from Theorem 3.9(2). See also [**110**].)

PROOF. We first prove part (1). If G has a directed cycle then the sum of the corresponding columns of D_G is zero. Conversely, if G does not have a directed cycle (that is, it is *acyclic*) then its vertices can be ordered so that every directed edge is directed towards the greater vertex. So, the following strict linear inequality is satisfied in $\mathcal{A}$: $\sum ix_i > 0$.

For parts (2) and (3) we give a full description of the subsets of $\mathcal{A}$ that span $n-1$ dimensional simplices. The corresponding subgraphs are necessarily spanning (all coordinates need to be used) and connected (otherwise a proper subset of coordinates has sum equal to zero) and have n elements. This makes each the union of a spanning tree with an extra edge. The converse is almost true: If a subgraph is a spanning tree plus one edge, then its corresponding subconfiguration is affinely independent (hence it spans an $(n-1)$-dimensional configuration) unless the cycle contained in the subgraph is balanced. So part (2) is clear.

If an unbalanced cycle does not exist, then there can not be an $(n-1)$-dimensional simplex. If there is an unbalanced cycle, it yields an $(n-1)$-dimensional simplex. Part (3) follows from the fact that the volume (with respect to the **A** lattice) of any such simplex is the difference in number of edges in each direction in its cycle. □

Hibi and Ohsugi also proved that if all induced cycles (cycles without a chord) satisfy part (3) (they are balanced or have one more edge in one direction) then $P_{\mathcal{A}}$ has a unimodular cover.

COROLLARY 3.15. (1) *If G is not acyclic then its arc polytope has regular unimodular triangulations.*

(2) *If G is bipartite with parts X and Y, and the edges are all directed from X to Y then all triangulations are unimodular.*

PROOF. For (1), a regular unimodular triangulation of $P_{\mathcal{A}}$ can be constructed by pulling the origin first and then the remaining points in any order. In such a triangulation, the origin is a vertex of every cell, so their lattice volumes are the determinants of the other vertices. Since the vertices are roots of $\mathbf{A}_{n-1}$, these will all be one.

In part (2), $\mathbf{0}$ is not in the affine hull of $P_{\mathcal{A}}$. So letting X and Y denote the two parts in G, we have $\mathcal{A} \subset \{\sum_{v\in X} x_a = -1\} \cap \{\sum_{v\in Y} x_a = 1\}$, and Theorem 3.9(2) implies the statement. □

3.2.2. Sub-configurations of $\mathbf{B}_n$, $\mathbf{C}_n$, $\mathbf{D}_n$ and $\mathbf{F}_4$. Much less is known about sub-configurations of other root systems.

THEOREM 3.16. *(Ohsugi and Hibi, [**101**] for part (1), [**100**] for part (2))*

(1) *$P_{\widetilde{\mathbf{B}}_n^+}, P_{\widetilde{\mathbf{C}}_n^+}$ and $P_{\widetilde{\mathbf{D}}_n^+}$ have* quadratic *triangulations.*

(2) *If $n \geq 2$, and $\mathcal{A}^+$ is a set of vectors satisfying the following conditions*
  (a) $\{\mathbf{e}_i + \mathbf{e}_j \mid 1 \leq i < j \leq n\} \subseteq \mathcal{A}^+ \subseteq \mathbf{B}_n^+ \cup \mathbf{C}_n^+$.
  (b) *$\mathcal{A}^+ \cap \mathbf{A}_{n-1}^+$ is transitively closed (that is, for all $1 \leq i < j < k \leq n$ with $\mathbf{e}_i - \mathbf{e}_j, \mathbf{e}_j - \mathbf{e}_k \in \mathcal{A}^+$ also $\mathbf{e}_i - \mathbf{e}_k \in \mathcal{A}^+$).*
  (c) *either all $\mathbf{e}_i \in \mathcal{A}^+$ or no $\mathbf{e}_i$ is in $\mathcal{A}^+$,*

then $P_{\widetilde{\mathcal{A}}^+}$ has a regular unimodular triangulation.

Producing triangulations for sub-configurations of (unions of) root systems is much more difficult if $\mathbf{0}$ is not a vertex. For $n \geq 6$ the toric ideal corresponding to $P_{\mathcal{A}}$ for $\mathcal{A} = \mathbf{A}_{n-1}^+, \mathbf{B}_n^+, \mathbf{C}_n^+, \mathbf{D}_n^+$ has no *quadratic* initial ideal, since it contains Example 3.13 as a face. So no regular unimodular triangulation can be *flag*. However, the ideal can still have a *square-free* initial ideal.

For $\mathbf{F}_4$ we know that some convex hulls of subsets of $\mathbf{F}_4$ are not integrally closed. For example, consider the polytope given by the convex hull P of the unit vectors together with $\mathbf{e}_1 + \mathbf{e}_2$. The point $\frac{1}{2}(1,1,1,1)$ is contained in $2P$, but is not a sum of two lattice points in P.

3.2.3. Edge Polytopes of Undirected Graphs. Sub-configurations of $\{\mathbf{e}_i + \mathbf{e}_j \mid 1 \leq i \leq j \leq n\} \subset \mathbf{C}_n^+$ have attracted some attention. They can be interpreted as *edge polytopes* of undirected graphs, perhaps with loops but without multiple edges. For this, let $G = (V, E)$ be a finite graph on $n = |V|$ vertices and $m = |E|$ edges. The *incidence matrix* $D_G = (d_{ve})$ of G is the $(n \times m)$ matrix with entries in $\{0, 1, 2\}$, where $d_{ve} = 1$ if e is incident to v, but not a loop, and $d_{ve} = 2$ if e is a loop at v, and all other entries of D_G are zero.

Letting $\mathcal{A}$ be the set of columns of D_G the *edge polytope* P_G of G is defined to be the convex hull of $\mathcal{A}$. Different graphs may define the same polytope. Namely, if there are loops attached to two vertices i and j of G, then adding or removing the edge between i and j does not change P_G. To avoid this ambiguity we define the graph $\widetilde{G}$ to be obtained from G by adding all edges between vertices incident to a loop. Then the only lattice points in P_G correspond to edges of $\widetilde{G}$.

Observe that Theorem 3.16 implies that the edge polytope P_{K_n} of the complete graph K_n has a regular unimodular triangulation.

LEMMA 3.17. *Let G be connected graph on n vertices.*

(1) *P_G is contained in the hyperplane $\sum x_i = 2$, and it affinely spans that hyperplane if and only if G has an odd cycle (that is, if it is not bipartite).*

(2) *If G is bipartite then P_G is totally unimodular.*

(3) *If G is not bipartite then a subgraph N with c connected components corresponds to the vertices of an $(n-1)$-dimensional simplex in P_G if and only if N spans all vertices and it has a unique cycle in each component, each of which is an odd cycle. In this case, the (lattice) volume of N is 2^{c-1}.*

PROOF. Part (2) follows from the fact that if G is bipartite with parts X and Y then orienting all edges from X to Y makes P_G (modulo a sign change on the coordinates corresponding to Y) the arc polytope of a directed bipartite graph satisfying the conditions of 3.9(2). Also, in this case P_A lies in the codimension-two affine subspace $\sum_{i\in X} x_i = \sum_{i\in Y} x_i = 1$, which is part of statement (1). For the rest of part (1), suppose that G is not bipartite. Then, it has an odd cycle C, and this cycle can be extended to a spanning subgraph H containing no other cycle. H has n vertices and n edges, and we only need to check that the determinant of the corresponding matrix is non-zero. In this determinant the rows and columns of vertices and edges that are not in the cycle can be neglected, and the determinant of the odd cycle itself is positive or negative one.

For part (3) note that, as in the directed case, the subgraph N, corresponding to a full dimensional simplex, needs to use all coordinates. However, it does not need to be connected. It cannot contain even cycles, since an even cycle produces an affine dependence (the alternating sum of the points corresponding to the edges in the cycle is zero). But it can contain odd cycles, as we saw in the proof of part (1). However, no two odd cycles can be in the same connected component. If two cycles have more than one vertex in common then there is an even cycle; If they have one or no vertices in common then, joining them by a path, if needed, yields an even *walk* in N, which also induces an affine dependence. So, we cannot have more than one cycle per component which means we cannot have more then n edges in total. Since we need n edges as the vertices of an $(n-1)$ simplex, we need all components to contain exactly one cycle, which proves the "only if" direction of (3).

For the converse, assume N is spanning and has a unique cycle in each component. Since N lies in an affine hyperplane not containing the origin, the (lattice) volume of its convex hull is the determinant of the corresponding matrix, divided by the (lattice) distance from that hyperplane to the origin. The latter is two. To show that the former is 2^c observe that the matrix has a block for each connected component, and the determinant of each block is positive or negative two, by the same argument as in part (1). □

The following result characterizes the existence of a unimodular cover of the edge polytope of a graph.

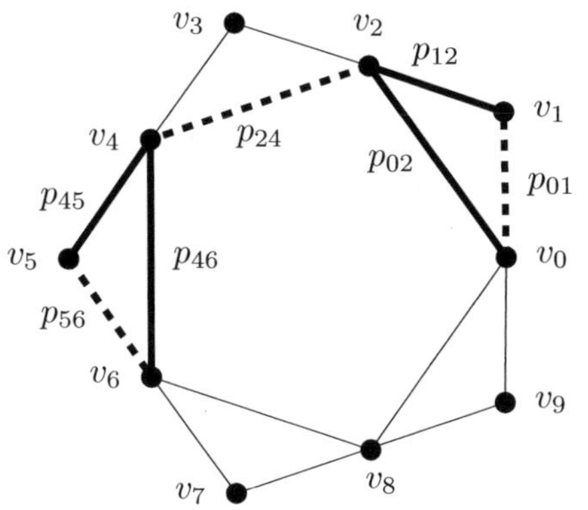

FIGURE 3.2. The graph G_{NRUT} of Example 3.19

THEOREM 3.18 (Ohsugi & Hibi '07 [**95, 98**], Simis, Vasconcelos, and Villarreal [**124**]). *Given a connected graph G, the polytope P_G has a unimodular cover if and only if no induced subgraph consists of two disjoint odd cycles.*

This characterization was originally conjectured by Simis, Vasconcelos, and Villarreal [**123**]. The condition of the graph G in the theorem is sometimes called the *odd cycle condition* (and usually stated in a different but equivalent way). In particular, this theorem implies that P_G is integrally closed if and only if $\widetilde{G}$ satisfies the odd cycle condition and that no simple graph violating the condition can have a regular unimodular triangulation.

PROOF. Every induced subgraph of G corresponds to the points in a particular face of G (the converse is not true). If an induced subgraph N consists of two disjoint odd cycles, then the corresponding face is a non-unimodular simplex, so P_G cannot have a unimodular cover.

Conversely, assuming the odd cycle condition, every disjoint pair of odd cycles in G can be connected to one another by a single edge, otherwise the minimal (induced) cycles contained in them violates the odd cycle condition.

To show that G_P has a unimodular cover it suffices to show that every non-unimodular simplex can be covered by simplices of smaller volume. For this, we show that if N is (the subgraph of G corresponding to) a non-unimodular simplex then we can add a single edge e to N so that all the simplices contained in $N \cup \{e\}$ have smaller volume than N.

So, let N be the subgraph corresponding to a non-unimodular simplex of full dimension. By the previous lemma, N has at least two connected components, each with an odd cycle. Let N_1 and N_2 be these cycles, and let e be an edge of G connecting them, which exists by the odd cycle condition. Then, $N \cup \{e\}$ has a unique affine dependence and it has, by Lemma 2.5, two triangulations, one using N as a simplex and the other not. In the affine dependence, e is the only point having coefficient positive or negative two. All others have coefficients of positive or negative one. This implies that N has twice the volume of every other simplex in $N \cup \{e\}$. □

However, not every polytope satisfying the odd-cycle condition has a regular unimodular triangulation. This was observed by Ohsugi and Hibi in [**97**].

EXAMPLE 3.19 ([**97**]). Let G_{NRUT} be the graph obtained from a 10-cycle with nodes numbered consecutively by $v_0, \ldots, v_9$ by adding the edges (v_0, v_2), (v_2, v_4), (v_4, v_6), (v_6, v_8) and (v_0, v_8). See Figure 3.2. This graph satisfies the odd cycle condition, so its edge polytope P_{NRUT} is integrally closed. It has a vertex p_{ab} corresponding to every edge (v_a, v_b) in the graph.

LEMMA 3.20. *P_{NRUT} does not have a regular unimodular triangulation.*

PROOF. The seven thick edges of Figure 3.2 form a five dimensional face F of P_{NRUT} (they are a non-bipartite induced subgraph with six vertices) with the unique affine dependence

$$p_{03} + p_{23} + p_{45} + p_{46} = p_{01} + 2p_{24} + p_{56}\,.$$

Since one of the sides has all coefficients equal to one but the other does not, only one of the two triangulations of F is unimodular (Lemma 2.6). If it is also regular, the weight vector defining it must choose the first term of the binomial $x_{03}x_{23}x_{45}x_{46} - x_{01}x_{24}^2x_{56}$ derived from the circuit (Theorem 2.22). That is, the weight vector $\boldsymbol{\omega} \in \mathbb{Z}^{15}$ must satisfy

$$\omega_{02} + \omega_{23} + \omega_{45} + \omega_{46} > \omega_{01} + 2\omega_{24} + \omega_{56}\,.$$

But there are four other faces obtained from this one by the pentagonal symmetry in G_{NRUT}, leading to the inequalities

$$\omega_{24} + \omega_{34} + \omega_{67} + \omega_{68} > \omega_{23} + 2\omega_{46} + \omega_{78}\,.$$
$$\omega_{46} + \omega_{56} + \omega_{08} + \omega_{89} > \omega_{45} + 2\omega_{68} + \omega_{01}\,.$$
$$\omega_{68} + \omega_{78} + \omega_{01} + \omega_{02} > \omega_{67} + 2\omega_{08} + \omega_{12}\,.$$
$$\omega_{08} + \omega_{09} + \omega_{23} + \omega_{24} > \omega_{89} + 2\omega_{02} + \omega_{34}\,.$$

Since the sum of the five left-hand sides equals the sum of the five right-hand sides, these five inequalities cannot be simultaneously satisfied. Therefore, P_{NRUT} cannot have a regular unimodular triangulation. □

It was shown by Firla and Ziegler [**51**] (and further studied by De Loera using his program PUNTOS) that the polytope P_{NRUT} of the previous example does have a non-regular unimodular triangulation. Later, Ohsugi [**94**] showed that G_{NRUT} can be generalized to an infinite family of integrally closed edge polytopes without regular unimodular triangulations. Each graph in this family is obtained by successively replacing a node of degree two with a path of length two.

The next theorem collects classes of graphs whose edge polytopes do have regular unimodular triangulations.

THEOREM 3.21. *Let $G = (V, E)$ be a finite simple graph, possibly with loops, and P_G the associated edge polytope.*

(1) *If G is bipartite, then all triangulations of P_G are unimodular. Further, P_G has a quadratic triangulation if and only if every minimal cycle in G has length four.*
(2) *If P_G is simple, but not a simplex, then P_G has a quadratic triangulation.*

(3) *If G does not contain a pair of disjoint odd cycles, then any regular triangulation of P_G is unimodular.*

PROOF. The first part of (1) follows from the fact that the incidence matrix of a bipartite graph is unimodular, and the convex hull of the rows does not contain $\mathbf{0}$. Necessity in the second part comes from the fact that an induced cycle of length $2k$ in G produces a face in P_G containing a unique circuit, and the circuit has k positive and k negative elements. Neither of the two triangulations of such a circuit is flag (Lemma 2.6). For sufficiency of the second part we refer to Hibi and Ohsugi in [**96**].

(2) follows from work of Hibi and Ohsugi [**102**, combining Prop. 1.1, Thms. 1.8 and 2.1]. In the same paper, they also characterize which graphs G give a simple edge polytope other than a simplex. Further, they prove that all such polytopes are smooth [**102**, Thm. 1.8]. Theorem 3.32 below provides an alternative proof of part (2) for the case where G has no loops.

(3) directly follows from Part 3 of Lemma 3.17. □

Our last result in this section is from [**97**], where Ohsugi characterizes which graph polytopes are compressed. We call G an *odd cycle graph* if G has at least two disjoint odd cycles and satisfies the odd cycle condition of Theorem 3.18. Given two disjoint odd cycles C and C' in a graph G we define

$$S_{C'}(C) := \left\{ v \in V(C) \;\middle|\; \begin{array}{c} v \text{ is incident to a chord of C} \\ \text{or a bridge between } C \text{ and } C' \end{array} \right\}.$$

$S_{C'}(C)$ decomposes C into paths. Let $s_{C'}(C)$ be the number of paths of odd length in this decomposition.

THEOREM 3.22 ([**97**, Thm. 4.1]). *P_G is compressed if and only if $s_{C'}(C) > 1$ for all disjoint pairs of odd cycles C, C' in G.*

Observe that the condition in the statement implies that G satisfies the odd cycle condition. Indeed, if G has an induced subgraph consisting of two cycles C and C', then $s_{C'}(C)$ equals one. The same paper also gives a complete characterization of all graphs G whose edge polytope has a regular unimodular pulling triangulation [**97**, Thm. 4.4]. Moreover, it asserts that for a simple odd cycle graph, the existence of a lifting function defining a regular unimodular triangulation can be tested by checking that a certain linearly defined subspace of edge space is non-empty. The linear inequalities only depend on bridges and chords in pairs of odd cycles [**97**, Thm. 3.5].

3.3. Other Graph Polytopes

3.3.1. Flow Polytopes. Let $G = (V, A)$ be a directed graph with n vertices and m edges. The incidence matrix D_G of G is the $(n \times m)$-matrix with a 1 at position (i, j) if edge j is directed towards i, -1 if j is directed away from i, and 0 otherwise. Let $\boldsymbol{a} \in \mathbb{Z}^n$ be a *demand vector* (i.e. some integer number for each vertex) and $\boldsymbol{l}, \boldsymbol{u} \in (\mathbb{Z} \cup \{\infty\})^m$ be upper and lower bounds for the flow on each edge. The *flow polytope* corresponding to the digraph D with demand $\boldsymbol{a}$ and bounds $\boldsymbol{l}, \boldsymbol{u}$ is

$$F_{D,\boldsymbol{a},\boldsymbol{l},\boldsymbol{u}} := \{\boldsymbol{x} \in \mathbb{R}^m \mid D_G \cdot \boldsymbol{x} = \boldsymbol{a},\ \boldsymbol{l} \leq \boldsymbol{x} \leq \boldsymbol{u}\}. \tag{3.1}$$

The dimension of $F_{D,\boldsymbol{a},\boldsymbol{l},\boldsymbol{u}}$ is at most $m-n+k$, where k is the number of connected components of the undirected graph corresponding to D. In the following we assume that $\sum_{i=1}^{n} a_i = 0$, as otherwise $F_{D,\boldsymbol{a},\boldsymbol{l},\boldsymbol{u}}$ is empty. Roughly, a flow polytope is the set of all assignments of a flow to all edges such that the upper and lower bounds are respected and at each vertex the incoming and outgoing flow differ by the demand at that vertex.

Transportation polytopes are the class of flow polytopes for which $D = \vec{\boldsymbol{K}}_{n,m}$, $\boldsymbol{a} = (-\boldsymbol{r}, \boldsymbol{c})$, $\boldsymbol{l} = \boldsymbol{0}$ and $\boldsymbol{u} = \infty$, where $\vec{\boldsymbol{K}}_{n,m}$ is the complete bipartite directed graph on m and n vertices. Fulkerson's integral flow theorem (see e.g. [**120**]) induces a path decomposition of any integral flow, which proves the following proposition.

PROPOSITION 3.23. *Flow polytopes are lattice polytopes.*

Birkhoff polytopes, denoted B_n, are another well known special case of transportation polytopes. They are the case where $\boldsymbol{r} = \boldsymbol{c} = \boldsymbol{1}$. B_n is an $(n-1)$-dimensional 0/1-polytope with facets $x_{ij} \geq 0$ for $1 \leq i, j \leq n$. This polytope can also be defined as the convex hull of all $(n \times n)$-permutation matrices or as the set of all $(n \times n)$-doubly stochastic matrices.

Consider the $n = 3$ case. Geometrically, the Birkhoff polytope B_3 is a direct sum of the two triangles. Viewed as the convex hull of permutation matrices, the two triangles are given by the odd and even permutations respectively. These triangles are not faces of B_3, and they form a circuit in B_3, i.e. a minimally dependent set. Hence, its toric ideal is the principal ideal

$$\mathcal{I}_{B_3} = \langle x_{123}x_{231}x_{312} - x_{132}x_{213}x_{321} \rangle.$$

$\mathcal{I}_{B_3}$ has two initial ideals, $\langle x_{123}x_{231}x_{312} \rangle$ and $\langle x_{132}x_{213}x_{321} \rangle$. This means $\mathcal{I}_{B_3}$ is not quadratically generated, and hence, B_3 does not have a quadratic triangulation. In fact, B_3 and its multiples are the only (3×3)-transportation polytopes that fail to have a quadratic triangulation.

THEOREM 3.24 (Haase, Paffenholz 07 [**61**, Thm. 1.5]). *If a (3×3)-transportation polytope $\mathrm{T}_{\boldsymbol{rc}}$ is not a multiple of B_3 then $\mathrm{T}_{\boldsymbol{rc}}$ has a quadratic triangulation.*

Ohsugi and Hibi showed that for $k \geq 2$ all multiples kB_3 have quadratic Gröbner bases [**103**], but their initial ideals are not square-free.

The proof of Theorem 3.24 uses pulling refinements of hyperplane subdivisions. Specifically, it looks at the set of hyperplanes obtained by fixing the variable corresponding to one edge in the definition of $F_{D,\boldsymbol{a},\boldsymbol{l},\boldsymbol{u}}$ (3.1). The transportation polytope is then subdivided by the hyperplanes corresponding to all but one pair of adjacent edges in $\vec{\boldsymbol{K}}_{n,m}$. The result is a finite list of lattice isomorphism types of dicing cells, and there is a linear functional that induces a flag pulling refinement on each cell. This gives a flag regular unimodular triangulation of the transportation polytope.

The following proposition shows that the class of transportation polytopes that are not multiples of some B_n includes all smooth transportation polytopes.

PROPOSITION 3.25 (Haase, Paffenholz 07 [**61**, Lemma 1]). *For an $(m \times n)$ transportation polytope $\mathrm{T}_{\boldsymbol{rc}}$, the following are equivalent.*

(1) $X_{\mathrm{T}_{\boldsymbol{rc}}}$ *is smooth.*
(2) $\mathrm{T}_{\boldsymbol{rc}}$ *is smooth.*
(3) $\mathrm{T}_{\boldsymbol{rc}}$ *is simple.*
(4) $\sum_{i\in I} r_i \neq \sum_{j \in J} c_j$ *for all index sets $I \subset [m]$, $J \subset [n]$ satisfying $|I| \cdot |J^c|, |I^c| \cdot |J| > 1$, where for $K \subset [n]$, K^c denotes the set $[n] \setminus K$.*

In his MSc-Thesis, M. Lenz [**82**] considered the case of smooth (3×4)-transportation polytopes with the following result.

PROPOSITION 3.26 (Lenz 07, [**82**, Satz 4.5.4]). *Every smooth (3×4)-transportation polytope has a quadratic triangulation.*

Based on experimental evidence, Diaconis and Eriksson conjectured that the toric ideal of the Birkhoff polytope is generated in degree three [**46**, Conj. 7]. This conjecture was later confirmed by Yamaguchi, Ogawa, and Takemura [**139**]. Their theorem leaves room for speculation about degrees of Gröbner bases and minimal non-faces of unimodular triangulations.

3.3.2. Polytopes defined by characteristic vectors of subsets of vertices and edges. Let $G = (V, E)$ be a finite simple graph with vertex set V and edge set E. For any subset $W \subset V$ the characteristic vector $\chi^W \in \mathbb{R}^V$ is given by $\chi_i^W = 1$ if $i \in W$ and 0 otherwise. For a subset $\mathcal{A} \subseteq 2^V$ the *characteristic polytope* is $P_{\mathcal{A}} := \mathrm{conv}(\chi_A \mid A \in \mathcal{A})$. We address such polytopes for various special choices of $\mathcal{A}$.

Given a graph $G = (V, E)$ *vertex cover* is a subset $C \subseteq E$ such that each $v \in V$ is incident to at least one $e \in C$. A vertex cover C is *minimal* if no proper subset of C is a vertex cover. Let $\mathcal{VC}$ denote the set of all minimal vertex covers.

THEOREM 3.27 (Herzog, Hibi, Ohsugi '09 [**64**]). *Let G be a bipartite graph. If all minimal vertex covers of G have the same size, then $P_{\mathcal{VC}}$ has a quadratic triangulation.*

A *stable set* in G is a subset S of the vertices such that no two vertices in S are connected by an edge. For the set $\mathcal{S}$ of all stable sets in G, the polytope $P_{\mathcal{S}}$ is commonly called the *stable set polytope* of G. A *clique* in a finite simple graph is an induced subgraph that is complete (i.e. every pair of vertices is connected by an edge). The *chromatic number* of a graph is the minimal number of colors needed to assign a color to each vertex so that no two adjacent vertices have the same color. Clearly, the chromatic number of a graph is at least the cardinality of the maximal clique in the graph. When the chromatic number of every subgraph in a finite simple graph equals the cardinality of its maximal clique, the graph is called *perfect*.

THEOREM 3.28. *The stable set polytope of a perfect graph is compressed.*

PROOF. Let P be the stable set polytope of a graph $G = (V, E)$. It is obvious that the following inequalities are feasible on P:

$$\begin{cases} x_v \geq 0 & \forall v \in V \\ \sum_{v \in C} x_v \leq 1 & \forall \text{ clique } C. \end{cases}$$

Chvátal [**38**, Thm. 3.1] proved that these inequalities actually define P if (and only if) G is perfect. Now, since clearly at every vertex of P we have

$$\begin{cases} x_v \leq 1 & \forall v \in V \\ \sum_{v \in C} x_v \geq 0 & \forall \text{ clique } C, \end{cases}$$

P has width one with respect to every facet. In particular, by Theorem 2.3, P is compressed. $\square$

As is somehow apparent in the above proof, for perfect graphs there is a duality between cliques (face inequalities) and stable sets (vertices). Since cliques in a graph G correspond to stable sets in the complement graph $\overline{G}$, [**38**, Thm. 3.1] is in fact a polyhedral proof of Lovász theorem that the complement of a perfect graph is perfect.

One example of perfect graph is the comparability graph of any poset $(X, \preccurlyeq)$. Its cliques and stable sets are, respectively, the chains and antichains of X. That is, the corresponding stable set polytope has as vertices the characteristic vectors of antichains and as defining inequalities $\sum_{i\in C} x_i \le 1$ for each chain C (together with $x_i \ge 0$ for every i). This polytope was studied by Stanley [**127**], who constructed a particular quadratic triangulation of it that piecewise linearly bijects to the dicing triangulation of the order polytope, mentioned in Section 3.1.1. As a corollary, the order polytope and the chain polytope of X have the same Ehrhart polynomial (in particular, the same volume, equal to the number of linear extensions of the poset).

REMARK 3.29. Order polytopes and chain polytopes have been generalized to *double posets* in [**36**]. Here, a double poset is a triple $(X, \preccurlyeq_1, \preccurlyeq_2)$ where $\preccurlyeq_1$ and $\preccurlyeq_2$ are two partial orders on X. The two orders, or the double poset, are said to be *compatible* if they have at least one common linear extension. The Cayley difference of two lattice polytopes $P_1, P_2 \subset \mathbb{R}^d$ is the lattice polytope $\operatorname{conv}(P_1 \times \{0\} \cup (-P_2) \times \{1\})$ (compare with the definitions of Cayley sum in Sections 2.3.3 and 4.3). With these preliminaries, the *double order polytope* (resp. *double chain polytope* of $(X, \preccurlyeq_1, \preccurlyeq_2)$ is the Cayley difference of the order polytopes (resp. of the chain polytopes) of $(X, \preccurlyeq_1)$ and $(X, \preccurlyeq_2)$.

Corollary 4.1 and Theorem 4.3 in [**36**] show that if $(X, \preccurlyeq_1, \preccurlyeq_2)$ is a compatible double poset then its double order and double chain polytopes have quadratic triangulations. Moreover, such triangulations can be constructed so that there is a piecewise linear map between them, as in the original case studied by Stanley. Corollary 4.7 in [**36**] gives a summation formula for the volume of these polytopes.

Going back to characteristic vectors related to graphs, an alternate common definition takes the set of all edges instead of the vertices as a base set, yielding vectors $\chi_W \in \mathbb{R}^E$. Then for any subset $U \subset V$, we can associate the subset $C_U \in E$ of all edges incident to exactly one node from U, and define the *cut polytope* $\mathsf{Cut}(G) \subset \mathbb{R}^E$ of G as the characteristic polytope given by the set $\mathcal{A} := \{C_U \mid U \subset V\}$.

THEOREM 3.30 (Sullivant 2004, [**134**]; see also [**133**]). *Let $G = (V, E)$ be a finite simple undirected graph. The cut polytope $\mathsf{Cut}(G)$ of G is compressed if and only if*

(1) *G has no K_5-minor, and*
(2) *all induced cycles in G have length at most four.*

In this context, G contains a K_5-minor if we can find five distinct vertices together with a set of pairwise internally vertex disjoint paths that connects each pair of the five vertices.

In discrete optimization one is often interested in whether the polytope associated to a combinatorial optimization problem is integrally closed, as this provides a way to solve integer linear optimization problems. In this setting, integrally closed polytopes are said to have the *integer decomposition property*. Examples include the s-t-connector polytope P_G of a directed graph $G = (V, A)$ (Trotter, see

[**121**, Thm. 13.8]), the base polytope, the spanning set polytope of a matroid and the independent set polytope of a matroid [**121**, Cor. 42.1e], and the up and down hull of the perfect matching polytope [**121**, Cor. 20.9c, 20.11b]. Note that this is not true for matching polytopes in general, as illustrated by the Petersen graph.

Various polytopes associated to combinatorial optimization problems are facet unimodular (i.e. the matrix of facet normals is totally unimodular). These include b-transhipment polytopes [**121**, Section 11.4], bipartite matching and bipartite perfect matching polytopes [**121**, Section 18.1], vertex cover [**121**, Section 18.4], edge cover and stable set polytopes of bipartite graphs [**121**, Section 19.5]. Hence, by Theorem 2.4 each of these classes have regular unimodular triangulations.

3.4. Lecture hall polytopes

Euler's classic result that there are as many partitions of an integer n into odd parts as there are partitions into distinct parts can be regarded as a "limit" ($d \to \infty$) of the following theorem by Bousquet-Mélou and Eriksson [**20**]: For every $n, d \in \mathbb{N}$ the number of so called d-lecture hall partitions of n is equal to the number of partitions of n into an odd and less than $2d$ number of parts. Here, a *d-lecture hall partition* is a partition $\lambda \in \mathbb{Z}^d$ satisfying the inequalities

$$0 \leq \frac{\lambda_1}{1} \leq \frac{\lambda_2}{2} \leq \ldots \leq \frac{\lambda_d}{d}.$$

As Savage says in her survey [**117**], "Over the past twenty years, lecture hall partitions have emerged as fundamental structures in combinatorics, number theory, algebra, and geometry, leading to new generalizations and interpretations of classical theorems and new results".

Already Bousquet-Mélou and Eriksson point out that the study of lecture hall partitions falls naturally within the theory of lattice points in cones [**21**, Sect. 5]. For $d \geq 1$, the lecture hall simplex is defined as

$$\mathrm{LHS}_{d+1} := \operatorname{conv} \begin{bmatrix} 0 & 0 & \cdots & \cdots & 0 & 1 \\ 0 & \vdots & & \iddots & 2 & 2 \\ 0 & \vdots & \iddots & \iddots & & \vdots \\ 0 & 0 & d-1 & \cdots & \cdots & d-1 \\ 0 & d & d & \cdots & \cdots & d \end{bmatrix} \subset \mathbb{R}^d.$$

It has the inequality description

$$\mathrm{LHS}_{d+1} = \left\{ \boldsymbol{x} \in \mathbb{R}^d \;\middle|\; 0 \leq \frac{x_1}{1} \leq \frac{x_2}{2} \leq \ldots \leq \frac{x_d}{d} \leq 1 \right\}.$$

We use $d+1$ rather than d as index since the cone over this simplex is unimodularly equivalent to the more familiar $d+1$st lecture hall cone. The triangulation in the following statement was communicated by C. Haase to the authors of [**13**], where it appears in detail. Beck, Braun, Köppe, Savage, and Zafeirakopoulos then ask [**13**, Conj. 6.1] whether the triangulation can be chosen in such a way that it elucidates the desirable enumerative properties of LHS_{d+1}.

THEOREM 3.31 ([**13**, Thm. 4.2]). *The lecture hall simplex* LHS_{d+1} *has a quadratic triangulation.*

PROOF. We proceed by induction on d. For $d = 1$, LHS_2 is a unit interval.

For $d \geq 2$, it is natural to intersect LHS_{d+1} with the hyperplane $x_d - x_{d-1} = 1$:

$$P := \{\boldsymbol{x} \in \mathrm{LHS}_{d+1} \mid x_d - x_{d-1} = 1\} = \operatorname{conv} \begin{bmatrix} 0 & \cdots & \cdots & 0 & 1 \\ \vdots & & \iddots & 2 & 2 \\ \vdots & \iddots & \iddots & & \vdots \\ 0 & d-1 & \cdots & \cdots & d-1 \\ 1 & d & \cdots & \cdots & d \end{bmatrix}$$

to obtain a simplex P which is unimodularly equivalent to LHS_d and thus has a quadratic triangulation $\mathcal{T}$ by induction. The hyperplane splits LHS_{d+1} into two simplices which can be triangulated in a compatible fashion.

$\underline{x_d - x_{d-1} \leq 1}$: This polytope below P is a lattice pyramid over P with apex the origin. Coning off $\mathcal{T}$, we obtain a quadratic triangulation which is compatible with any weights inducing $\mathcal{T}$ on P.

$\underline{x_d - x_{d-1} \geq 1}$: This polytope above P is (equivalent to) a chimney polytope over P (compare Sect. 2.2). In fact, let $\pi\colon \mathbb{R}^d \to \mathbb{R}^{d-1}$ be the projection that forgets the last coordinate, and let $P' := \pi(P) \cong P$. Then

$$\{\boldsymbol{x} \in \mathrm{LHS}_{d+1} \mid x_d - x_{d-1} \geq 1\} = \{(\boldsymbol{y}, y_d) \in P' \times \mathbb{R} \mid y_{d-1} + 1 \leq y_d \leq d\}.$$

Thus, we can lift $\mathcal{T}$ to a quadratic triangulation of the polytope above P. □

More generally, for any sequence $\boldsymbol{s} = (s_i)_{i=1}^d$, Savage and Schuster [**118**] define the $\boldsymbol{s}$-lecture hall simplex $\mathrm{LHS}_{d+1}(\boldsymbol{s})$ to be

$$\mathrm{LHS}_{d+1}(\boldsymbol{s}) := \left\{ \boldsymbol{x} \in \mathbb{R}^d \;\middle|\; 0 \leq \frac{x_1}{s_1} \leq \frac{x_2}{s_2} \leq \cdots \leq \frac{x_d}{s_d} \leq 1 \right\}. \tag{3.2}$$

Equivalently,

$$\mathrm{LHS}_{d+1}(\boldsymbol{s}) := \operatorname{conv} \begin{bmatrix} 0 & \cdots & 0 & s_1 \\ \vdots & \iddots & \iddots & \vdots \\ 0 & s_d & \cdots & s_d \end{bmatrix} \subset \mathbb{R}^d.$$

It would be desirable to understand for which $\boldsymbol{s}$ this simplex has a unimodular triangulation.

The simplices $\mathrm{LHS}_{d+1}(\boldsymbol{s})$ appear also in the following works: (1) The fundamental parallelepiped of the simplex $\mathrm{LHS}_{d+1}(\boldsymbol{s})$ has been studied by Liu and Stanley [**83**]. (2) Consider a finite partial order $(X, \preccurlyeq)$ as in Section 3.1.1, and take as additional input a function $\boldsymbol{s} \in \mathbb{Z}_{>0}^X$ giving positive integer weights to its elements. Then Brändén and Leander [**22**] define an $\boldsymbol{s}$-order polytope $\mathcal{O}(\preccurlyeq, \boldsymbol{s})$ as the image of the order polytope $\mathcal{O}(\preccurlyeq)$ under the coordinate-wise scaling $x \mapsto (s_1x_1, \ldots, s_nx_n)$. Via this map, the canonical triangulation of the usual order polytope triangulates $\mathcal{O}(\preccurlyeq, \boldsymbol{s})$ into the simplices $\mathrm{LHS}_{n+1}(\sigma(\boldsymbol{s}))$ where the permutation σ runs over all linear extensions of $\preccurlyeq$.

3.5. Smooth Polytopes

We now turn our attention to two classes of smooth polytopes, those with no interior points and those with one interior lattice point, satisfying a special condition.

3.5.1. Empty Polytopes. First we consider lattice polytopes whose only lattice points are their vertices. The following theorem establishes strong restrictions on the combinatorics of such polytope if they are smooth.

THEOREM 3.32. *Every smooth polytope P such that $P \cap \mathbb{Z}^d = \mathcal{V}(P)$ is lattice equivalent to a product of unimodular simplices.*

Together with Proposition 2.11, this yields the following corollary.

COROLLARY 3.33. *Any smooth lattice polytope P satisfying $P \cap \mathbb{Z}^d = \mathcal{V}(P)$ has a quadratic triangulation.*

The proof of the theorem is very much inspired by Kaibel and Wolff's proof of a slightly different result.

THEOREM 3.34 (Kaibel, Wolff '00 [**70**]). *Any simple $0/1$-polytope is the product of some $0/1$-simplices.*

Note that in the vertex-edge graph of a simple polytope every vertex has d neighbors, and for any choice of neighbors N of a vertex $v \in P$ in the graph of P, there is a unique face F of P such that $\{v\} \cup N \subseteq F$ and N is the set of neighbors of v in the graph of F. This face is denoted $F(v, N)$. The following lemma is needed for our proof of Theorem 3.32. The crucial observation in both proofs is that every face of P is again smooth and all its lattice points are vertices. In particular, every two-face is either a standard triangle or a unit square.

LEMMA 3.35. *If P is a smooth lattice polytope such that $P \cap \mathbb{Z}^d = \mathcal{V}(P)$, $v \in \mathcal{V}(V)$, and N is a set of neighbors of v in the graph of P such that no two vertices in N are adjacent in the graph of P, then $F(v, N)$ is a combinatorial cube.*

PROOF. Up to a unimodular transformation we can assume that $\boldsymbol{v}$ is the origin $\mathbf{0}$ and the adjacent vertices are $\boldsymbol{e}_1, \dots, \boldsymbol{e}_d$.

Let $E_I := \{\boldsymbol{e}_i \mid i \in I\}$ for some $I \subset \{1, \dots, r\}$ be a set of non-adjacent vertices. We use induction on the coordinate sum $\boldsymbol{v}_I := \sum_{i \in I} \boldsymbol{e}_i$. If $|I| = 2$, then by the above observation the vertices in E_I span a square at $\mathbf{0}$. Hence, $\boldsymbol{v}_I \in P$. If $|I| \geq 3$, then by induction all partial sums of the elements of E_I are contained in P. Hence, they span a face that differs from the cube by at most the vertex $\boldsymbol{v}_I$. But in dimensions three and higher, the cube minus a vertex is not a smooth polytope, so $\boldsymbol{v}_I \in P$. □

PROOF OF THEOREM 3.32. We can again assume that $\mathbf{0}$ is a vertex with neighbors $\boldsymbol{e}_i$ $(1 \leq i \leq d)$. So the incident facet inequalities are $x_i \geq 0$ $(1 \leq i \leq d)$. Since P is a simple polytope, any k of the $\boldsymbol{e}_i$ define a k-dimensional face.

The first step is to show that "being adjacent" is an equivalence relation among the vertices $\boldsymbol{e}_i$. That is, if for pairwise distinct $1 \leq i, j, k \leq d$, $\boldsymbol{e}_i$ is adjacent to $\boldsymbol{e}_j$, and $\boldsymbol{e}_j$ is adjacent to $\boldsymbol{e}_k$, then all three are contained in a common two-dimensional face. Suppose $\boldsymbol{e}_i$ and $\boldsymbol{e}_k$ were not adjacent, then $\boldsymbol{e}_i - \boldsymbol{e}_j + \boldsymbol{e}_k$ would be a forth vertex in this face. But this violates the inequality $x_j \geq 0$. So, $\boldsymbol{e}_i$ and $\boldsymbol{e}_k$ are adjacent, and the three vertices are in a common two dimensional face.

If $C_0, \dots, C_r$ are the equivalence classes of this relation, then for $1 \leq j \leq r$, $C_j \cup \{\mathbf{0}\}$ spans a unimodular simplex Δ_j. Letting $Q := \Delta_1 \times \cdots \times \Delta_r$ be the product of these simplices, we aim to show that $P = Q$.

By the previous lemma all vertices of Q are contained in P. It remains to show that all edges of Q are edges of P. Before proving this we show that this will finish the proof. The graph of P is connected, so if there was a vertex v of P that was not a vertex of Q, then at least one of the vertices of P that was in Q would have a neighbor in P that was not a neighbor in Q, so P would not be simple.

We now show that all edges of Q are edges of P. By the previous lemma, if E_I is a set of pairwise non-adjacent neighbors of $\mathbf{0}$, then all edges of the cube $F(\mathbf{0}, E_I)$ are in P. It remains to argue that if $\boldsymbol{e}_1$ and $\boldsymbol{e}_2$ are adjacent and $\{1\} \sqcup I$ indexes pair-wise non-adjacent vertices of P, then $\boldsymbol{e}_1 + \boldsymbol{e}_I$ and $\boldsymbol{e}_2 + \boldsymbol{e}_I$ are adjacent. Assume they are not adjacent. The faces $F(\mathbf{0}, \{\boldsymbol{e}_1\} \cup E_I)$ and $F(\mathbf{0}, \{\boldsymbol{e}_2\} \cup E_I)$ are cubes. So as neighbors of the vertex v_I, vertices $\boldsymbol{e}_1 + \boldsymbol{v}_I$, $\boldsymbol{e}_2 + \boldsymbol{v}_I$, and $\boldsymbol{v}_I - \boldsymbol{e}_i$ are pair-wise non-adjacent, for $i \in I$. Hence, they span a cube at $\boldsymbol{v}_I$, contradicting the adjacency of $\boldsymbol{e}_1$ and $\boldsymbol{e}_2$. □

3.5.2. Reflexive Polytopes. In this section we report on a class of polytopes where a computational approach implementing pull-back and push-forward subdivisions has been quite successful.

A lattice polytope P is called *reflexive*, if it contains a unique interior lattice point and all facets are lattice distance one from this point. Without lose of generality it can be assumed that the interior point is the origin. The *polar* of a polytope $P \subseteq \mathbb{R}^d$ with $\mathbf{0} \in \operatorname{int}(P)$ is

$$P^\vee := \{\boldsymbol{u} \in \mathbb{R}^d \mid \langle \boldsymbol{x}, \boldsymbol{u} \rangle \geq -1 \,\forall \boldsymbol{x} \in P\}.$$

If P is reflexive then $P^\vee$ is again a lattice polytope. P is *smooth* if P is simple and the primitive generators of every vertex cone span the lattice.

THEOREM 3.36. *For $d \leq 3$, every reflexive d-dimensional lattice polytope has a regular unimodular triangulation.*

PROOF. The origin is the unique interior lattice point of P. Any pulling triangulation in which the origin is pulled first will be regular and unimodular. For $d = 3$ we use the fact that any full triangulation of a polygon is unimodular. □

There are $5, 18, 124, 866, 7622$, 72256, 749892, and 8229721 *smooth* reflexive polytopes in dimensions two, three, four, five, six, seven, eight, and nine respectively (up to lattice equivalence). Those of dimension up to eight were first computed by Øbro [**92**] and a full list of explicit representatives can be obtained from the `polymake database` (www.polymake.org/doku.php/data); the classification has been extended to dimension nine by Lorenz and the third author, see polymake.org/polytopes/paffenholz/www/fano.html.

We used two approaches to establish regular unimodular triangulations of the polars of these polytopes up to dimension seven (checking dimensions eight and nine is in progress):

(1) We checked whether the facet normals define a unimodular system (then the existence of a triangulation follows from Theorem 2.4).
(2) We searched for a sequence of projections along coordinate directions satisfying the conditions for push-forward and pull-back subdivisions given in Section 2.2. If such a sequence projects the polytope down to dimension two, the polytope has a regular unimodular triangulation (as all two-dimensional lattice polytopes have regular unimodular triangulations). If there is such a sequence projecting the polytope to dimension one, we know that the polytope has a *quadratic* triangulation.

dim.	number of polytopes	RUT	quadratic triangulation	facet unimodular	projects to dimension one
2	5	5	5	5	5
3	18	18	18	16	18
4	124	124	124	96	124
5	866	866	866	554	866
6	7622	7622	$\geq$7620	4097	$\geq$7620
7	72256	72256	$\geq$72240	31881	$\geq$72240

TABLE 3.1. Number of smooth reflexive polytopes of dimension$\leq$ 7; all of them possess regular unimodular triangulations, and all except perhaps 18 possess quadratic ones.

For those polytopes for which a two-dimensional, but no one-dimensional projections were found, individual inspection confirmed a quadratic triangulation for some of them.

Both checks were done with software system `polymake` [**69**, **109**] using an extension for projections of polytopal subdivisions[**104**]. The detailed results are listed in Table 3.1, and are summarized in the next theorem.

THEOREM 3.37 (Haase, Paffenholz '09 [**60**]).
- *All smooth reflexive d-polytopes for $d \leq 8$ are integrally closed.*
- *All smooth reflexive polytopes in dimension six have a regular unimodular triangulation, and all but at most two have a quadratic triangulation.*
- *All smooth reflexive polytopes in dimension seven have a regular unimodular triangulation, and all but at most* 16 *have a quadratic triangulation.*

The remaining two smooth Fano polytopes in dimension six and the 16 smooth Fano polytopes in dimension seven may still have a quadratic triangulation, but we were not able to construct such a triangulation with our approach. We checked whether the polytopes are integrally closed with the `polymake` interface to `Normaliz` [**31**]. Previously it was shown by Piechnik that all smooth reflexive d-polytopes have a regular unimodular triangulation for $d \leq 4$. Computations in dimensions eight and nine are currently work in progress. Data vor these computation can be found at `polymake.org/polytopes/paffenholz/www/rut.html`.

3.6. The Gröbner fan and the toric Hilbert scheme

3.6.1. The Gröbner fan and the secondary fan. Here we examine the relation between Gröbner bases and subdivisions, mentioned in section 2.4.

As in section 2.4,[1] we let $\mathcal{A} := (P \times \{1\}) \cap \mathbb{Z}^{d+1}$ be the homogenized set of lattice points in a polytope $P \in \mathbb{R}^d$, take the polynomial ring $S := \Bbbk[x_{\boldsymbol{a}} : \boldsymbol{a} \in \mathcal{A}]$ with one variable for each lattice point in P, and consider the toric ideal of P, i.e., the binomial ideal generated by linear dependences among the lattice points:

$$\mathcal{I}_P = \left\langle \boldsymbol{x}^{\boldsymbol{m}} - \boldsymbol{x}^{\boldsymbol{n}} \ : \ \boldsymbol{m}, \boldsymbol{n} \in \mathbb{Z}^{\mathcal{A}}_{\geq 0} \ , \ \sum_{\boldsymbol{a} \in \mathcal{A}} m_{\boldsymbol{a}} \boldsymbol{a} = \sum_{\boldsymbol{a} \in \mathcal{A}} n_{\boldsymbol{a}} \boldsymbol{a} \right\rangle .$$

[1]Everything we say in this section can be extended to the more general case where $\mathcal{A}$ is any finite and homogeneous set of lattice vectors in $\mathbb{Z}^{d+1}$

Recall, each choice of weights $\boldsymbol{\omega} \in \mathbb{R}^{\mathcal{A}}$ induces a regular subdivision $\mathcal{T}_{\boldsymbol{\omega}}$ of P, on the lattice polytope side, and an initial ideal $\mathrm{in}_{\boldsymbol{\omega}} \mathfrak{I}_P := \langle \mathrm{in}_{\boldsymbol{\omega}} f \ : \ f \in I \rangle$, on the toric algebra side.

However, if $\boldsymbol{\omega}$ is not *generic*, the subdivision $\mathcal{T}_{\boldsymbol{\omega}}$ may not be a triangulation, and the ideal $\mathrm{in}_{\boldsymbol{\omega}} \mathfrak{I}_P$ may not be a monomial. This means a polynomial f may have several monomials of highest weight with respect to $\boldsymbol{\omega}$, and all those monomials form the leading part of f, which we also denote $\mathrm{in}\, f$. For example, if $\boldsymbol{\omega}_{\boldsymbol{a}}$ is the same constant for every $\boldsymbol{a}$ then:

- $\mathcal{T}_v \boldsymbol{\omega}$ is the *trivial subdivision* (P itself is its only full-dimensional cell) and
- $\mathrm{in}_{\boldsymbol{\omega}} \mathfrak{I}_P = \mathfrak{I}_P$, since $\mathfrak{I}_P$ is homogeneous and for a homogeneous f, and $\mathrm{in}\, f = f$.

We state without proof two properties relating the initial ideals $\mathrm{in}_{\boldsymbol{\omega}} \mathfrak{I}_P$ and the corresponding regular subdivisions $\mathcal{T}_{\boldsymbol{\omega}}$. (See [**132**, Chapter 10] for details.)

THEOREM 3.38 (Sturmfels 96 [**132**]). *If* $\mathrm{in}_{\boldsymbol{\omega}_1} \mathfrak{I}_P = \mathrm{in}_{\boldsymbol{\omega}}(\mathrm{in}_{\boldsymbol{\omega}_2} \mathfrak{I}_P)$ *for some* $\boldsymbol{\omega}_1$, $\boldsymbol{\omega}_2$ *and* $\boldsymbol{\omega}$ *(that is, if* $\mathrm{in}_{\boldsymbol{\omega}_1} \mathfrak{I}_P$ *is an initial ideal of* $\mathrm{in}_{\boldsymbol{\omega}_2} \mathfrak{I}_P$*) then the subdivision* $\mathcal{T}_{\boldsymbol{\omega}_1}$ *refines the subdivision* $\mathcal{T}_{\boldsymbol{\omega}_2}$.

COROLLARY 3.39. *If* $\mathrm{in}_{\boldsymbol{\omega}_1} \mathfrak{I}_P = \mathrm{in}_{\boldsymbol{\omega}_2} \mathfrak{I}_P$ *for different* $\boldsymbol{\omega}_1$ *and* $\boldsymbol{\omega}_2$, *then* $\mathcal{T}_{\boldsymbol{\omega}_1} = \mathcal{T}_{\boldsymbol{\omega}_2}$. *This follows from the previous property by letting* $\boldsymbol{\omega} = \boldsymbol{\omega}_1$ *(and then switching the roles of* $\boldsymbol{\omega}_1$ *and* $\boldsymbol{\omega}_2$*).*

These properties mean that there is an order-preserving map from the poset of all initial ideals of $\mathfrak{I}_P$ to the poset of all regular subdivisions of $\mathcal{A}$, where the latter are partially ordered by refinement and the former are ordered by $\mathfrak{I}_1 < \mathfrak{I}_2$ if $\mathfrak{I}_1$ is an initial ideal of $\mathfrak{I}_2$. This was proved by Sturmfels in [**130**].

Put another way, the regular subdivision and the initial ideal construction provide two stratifications of the vector space $\mathbb{R}^{\mathcal{A}}$. The first is based sets of $\boldsymbol{\omega}$'s which give the same regular subdivisions and the second is based on which $\boldsymbol{\omega}$'s give the same initial ideals. Both stratifications are complete rational polyhedral fans, and they are called , respectively, the *secondary fan* and the *Gröbner fan* of $\mathcal{A}$.

THEOREM 3.40 (Sturmfels 91, [**130**]). *The Gröbner fan of* $\mathcal{A}$ *refines the secondary fan of* $\mathcal{A}$.

Observe that for the secondary fan to be well defined, the subdivisions of $\mathcal{A}$ need to be considered as subdivisions of point configurations, as when we defined weak pulling in Section 2.1.1.

Theorem 3.40 implies that $\mathcal{T}_\omega$ can be clearly determined from $\mathrm{in}_\omega \mathfrak{I}_P$.

3.6.2. The toric Hilbert scheme. We now look at a property that is shared by all initial ideals of $\mathfrak{I}_P$ and the toric ideal $\mathfrak{I}_P$ itself.

As detailed in [**132**, Chapter 10], $\mathcal{A}$ defines a d-dimensional multi-grading on the polynomial ring $\Bbbk[x_{\boldsymbol{a}} : \boldsymbol{a} \in \mathcal{A}]$, assigning multi-degree $\boldsymbol{a}_i$ to the variable x_i. Ideals $\mathfrak{I} \subset \Bbbk[x_{\boldsymbol{a}} : \boldsymbol{a} \in \mathcal{A}]$ that are homogeneous with respect to this grading have well-defined Hilbert functions

$$\begin{array}{ccc} \mathbb{Z}_{\geq 0}\mathcal{A} & \longrightarrow & \mathbb{Z}_{\geq 0} \\ \boldsymbol{b} & \longmapsto & \dim_{\Bbbk} \mathfrak{I}_{\boldsymbol{b}} \end{array}$$

where $\mathbb{Z}_{\geq 0}$ is the set of non-negative integers, $\mathbb{Z}_{\geq 0}\mathcal{A}$ is the semigroup of non-negative integer combinations of $\mathcal{A}$, and for each $\boldsymbol{b} \in \mathbb{Z}_{\geq 0}\mathcal{A}$, $\mathfrak{I}_{\boldsymbol{b}}$ is the degree $\boldsymbol{b}$ part of $\mathfrak{I}$.

The most natural $\mathcal{A}$-homogeneous ideal is the toric ideal $\mathcal{I}_P$, generated by the binomials

$$\{ \boldsymbol{x}^{\boldsymbol{m}} - \boldsymbol{x}^{\boldsymbol{n}} \ : \ \boldsymbol{m}, \boldsymbol{n} \in \mathbb{Z}^{\mathcal{A}}_{\geq 0} \ , \ \sum_{\boldsymbol{a}\in\mathcal{A}} m_{\boldsymbol{a}}\boldsymbol{a} = \sum_{\boldsymbol{a}\in\mathcal{A}} n_{\boldsymbol{a}}\boldsymbol{a} \ \},$$

because every $\boldsymbol{b} \in \mathbb{Z}_{\geq 0}\mathcal{A}$, $(\mathcal{I}_P)_{\boldsymbol{b}}$ has codimension one in $(\Bbbk[x_{\boldsymbol{a}} : \boldsymbol{a} \in \mathcal{A}])_{\boldsymbol{b}}$. This characterizes the Hilbert function of $\mathcal{I}_{\boldsymbol{b}}$.

An $\mathcal{A}$-homogeneous ideal $\mathcal{I} \subset \Bbbk[x_{\boldsymbol{a}} : \boldsymbol{a} \in \mathcal{A}]$ is called *$\mathcal{A}$-graded* if it has the same Hilbert function as the toric ideal $\mathcal{I}_P$. $\mathcal{A}$-graded ideals include all the initial ideals of $\mathcal{I}_P$, but can include other ideals as well. The *toric Hilbert scheme*, as introduced by Peeva and Stillman [**107**], is the set of all $\mathcal{A}$-graded ideals with a suitable algebraic structure defined by some determinental equations. An equivalent description via binomial equations appeared in [**131**, §6]. (See also [**85, 129**].)

Surprisingly, the $\mathcal{A}$-graded ideals that are not initial are still related to subdivisions of $\mathcal{A}$. Sturmfels [**132**, Theorem 10.10] proved that the order-preserving map implied in Theorem 3.40 extends to an order preserving map from the poset of all $\mathcal{A}$-graded ideals, where the partial ordered is given by by "toric deformation" (a generalization of the property of being an initial ideal), to the poset of all subdivisions, still ordered by refinement.

That is to say, every $\mathcal{A}$-graded ideal I has a canonically associated polyhedral subdivision $\mathcal{T}_I$ of $\mathcal{A}$. If I is monomial, then $\mathcal{T}_I$ is a triangulation, whose simplices are spanned by the standard monomials in $\Bbbk[x_{\boldsymbol{a}} : \boldsymbol{a} \in \mathcal{A}]/\operatorname{Rad}(I)$.

Santos [**115**] used this map to show the existence of non-connected toric Hilbert schemes, elaborating on work of Maclagan and Thomas [**85**]. Here are the main ideas. There is a natural and well-known adjacency relation between triangulations of the same configuration. It can be defined as an operation that takes out certain simplices and inserts others, but it is equivalent to the following [**44**, Section 2.4]: two triangulations $\mathcal{T}_1$ and $\mathcal{T}_2$ of $\mathcal{A}$ are related by a *geometric bistellar flip* (or just *flip*, for short) if there is a polyhedral subdivision $\mathcal{T}$ of $\mathcal{A}$ whose only refinements are $\mathcal{T}_1$ and $\mathcal{T}_2$. Maclagan and Thomas defined an analogous adjacency relation between $\mathcal{A}$-graded monomial ideals (*mono-$\mathcal{A}$-GIs* for short), which they also called flip. Their relation has the following properties.

PROPOSITION 3.41 (MacLagan & Thomas 02, [**85**]). (1) *A toric Hilbert scheme is connected if and only if the graph of mono-$\mathcal{A}$-GIs is connected.*
(2) *Triangulations of $\mathcal{A}$ corresponding to neighboring mono-$\mathcal{A}$-GIs either coincide or differ by a geometric bistellar flip.*

Lattice point configurations with non-connected graphs of flips are rare, but they were shown to exist in 2000 [**114**]. Still, this does not necessarily imply the corresponding toric Hilbert scheme is non-connected, because the Sturmfels map going from $\mathcal{A}$-graded ideals to subdivisions of $\mathcal{A}$ is in general not surjective [**132**, Example 10.13], [**107**]. However, Maclagan and Thomas also observed, based on [**132**, Theorem 10.14], that the image of the map always contains all the unimodular triangulations of $\mathcal{A}$.

COROLLARY 3.42. *The toric Hilbert scheme of $\mathcal{A}$ has at least as many connected components as there are connected components in the graph of triangulations of $\mathcal{A}$ that contain unimodular triangulations.*

Santos' non-connected toric Hilbert scheme is based on the construction of a polytope P for which the corresponding configuration $\mathcal{A}$ has unimodular triangulations in different components of the graph of flips. Let $Q \in \mathbb{R}^4$ be the 24-cell. The 24-cell is one of the six regular four-dimensional polytopes, and can be realized as the convex hull of the root system of type $\mathbf{D}_4$. We consider it as a lattice polytope in the lattice it generates, $D_4 := \{(a_1, a_2, a_3, a_4) \in \mathbb{Z}^4 : \sum a_i \in 2\mathbb{Z}\}$. Its only lattice points are the origin and the 24 roots $\pm \boldsymbol{e}_i \pm \boldsymbol{e}_j$, with $(i,j) \in \binom{[4]}{2}$. Q has 24-facets, all regular octahedra, and 96 two-faces, all regular triangles.

THEOREM 3.43 (Santos 05, [**115**], see also [**44**, Section 7.3]). *If $P = Q \times [0,1]$ is a lattice polytope in $D_4 \times \mathbb{Z}$, such that $\mathcal{A}$ has 50 points: the 48 vertices of P plus the centers of $Q \times \{0\}$ and $Q \times \{1\}$, then:*

(1) *The graph of triangulations of $\mathcal{A}$ has at least 13 connected components, each containing at least 3^{48} unimodular triangulations.*
(2) *The toric Hilbert scheme of $\mathcal{A}$ has at least 13 connected components each containing at least 3^{48} monomial ideals, and they all have dimension at least 96.*

PROOF. The proof uses the idea of *staircase refinements* introduced in Section 2.3. We sketch it here, and refer to [**44**, Section 7.3] or [**115**] for details. The crucial tool is an orientation of the graph of the two-cell with the following properties:

- It is locally acyclic, but no edge can be reversed in it without creating a cycle in one of the 96 triangular faces of D_4
- In each of the 24 octahedral facets of Q, one (and only one) of the four-cycles of edges is a directed cycle.

Figure 3.3 depicts this orientation, in several pieces. To understand the figure, observe that that 24 vertices of Q are the union of the 16 vertices of a four-cube plus the eight points $\pm 2e_i$. Moreover, the graph of Q is the graph of the four-cube plus the edges joining each of the eight three-cube facets of it to the corresponding e_i or $-e_i$. The figure shows stereographic projections of the eight three-cube facets of the four-cube, each with the orientation of its edges. The projection makes two of the facets regular cubes, so that one of the extra points is sent to infinity. We leave it to the reader to check that the orientation displayed has the two claimed properties (thanks to symmetry this task is not as hard as it might seem).

In order to triangulate P we first triangulate each octahedral facet of Q in the unique way that does not create a local cycle. That is, in each octahedron use the diagonal that is orthogonal to the quadrilateral cycle of that octahedron. Also, by the properties of the triangulation, this new diagonal can only be given one of its two orientations without creating a local cycle. Cone this triangulation of the boundary of Q to the origin, and orient every new edge away from the origin. This produces a triangulation $\mathcal{T}$ of Q with a locally acyclic orientation. Further, this triangulation and orientation is unimodular, and it has no reversible edge. That is, reversal of any individual edge creates a directed cycle. For the 96 original edges of the 24-cell, this is part of what we claimed before. It is also true and easy to check for the new 24 diagonals of the octahedra and 24 edges from the center.

We can now triangulate P using the staircase refinement of $\mathcal{T}$ and the edge $[0,1]$ (we give to $[0,1]$ either of its two orientations). It turns out that the fact that no edge can be reversed in the locally acyclic orientation implies that this

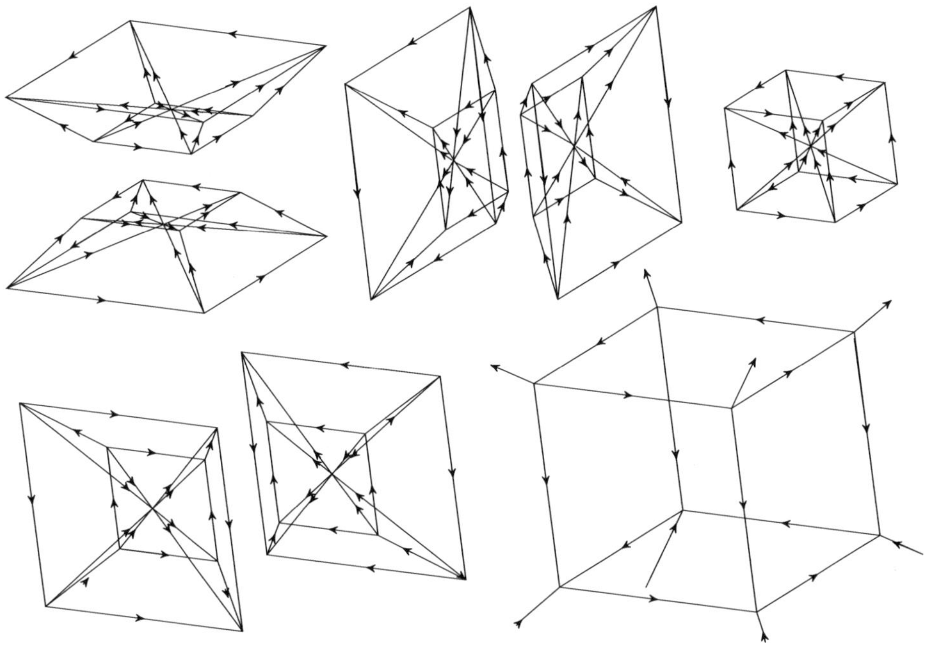

FIGURE 3.3. A locally acyclic orientation of the graph of a 24-cell.

triangulation cannot be connected by flips to any triangulation constructed in the same way using a different initial orientation of the edges of the 24-cell. This proves that the graph of triangulations contains unimodular triangulations in different components. There are at least 13 such components because symmetries of the 24-cell produce 12 different ways of constructing the initial locally acyclic orientation, plus another component that will contain all the *globally acyclic* orientations. The 3^{48} triangulations in each component and the 96 in the dimension of the toric Hilbert scheme are by-products of the many symmetries of the construction. □

CHAPTER 4

Dilations and the KMW Theorem

Dilating a lattice polytope by a positive integer is a natural operation. One of the first theorems about unimodular triangulations was proved in the early days of toric geometry by Knudsen, Mumford, and Waterman[1] [**77**], who were interested in semi-stable reduction of families over curves.

THEOREM 4.1 (KMW theorem, [**77**]). *Given a polytope P, there is a $c \in \mathbb{Z}_{>0}$ such that the dilation $c \cdot P$ admits a regular unimodular triangulation.*

We say that c is a *KMW-number* of P if cP has a unimodular triangulation. This KMW-theorem raised several questions that are still open, including:

- What is the minimum $c(P)$ for a given polytope P? Is there a $c(d)$ that is a KMW-number for every polytope of dimension d?
- What is the structure of the set of KMW-numbers of a given P? Is it a monoid? Theorem 4.8 implies it is closed under taking multiples of an element. It is not clear whether it is closed under taking sums.
- To our knowledge, Example 4.13 below is the first class of polytopes P and integers c in the literature such that c is a KMW-number for P but $c+1$ is not.

The following two theorems by Bruns, Gubeladze, and Trung; and Eisenbud, Reeves, and Totaro, respectively, show that we cannot expect algebraic obstructions to these questions.

THEOREM 4.2 ([**28**]). *$c \cdot P$ is integrally closed for $c \geq d-1$, and Koszul for $c \geq d$.*

THEOREM 4.3 ([**48**]). *After a linear change of coordinates, $\mathfrak{I}_{cP}$ has a quadratic Gröbner basis for $c \geq \operatorname{reg}(\mathfrak{I}_P)/2$.*

In this respect if we relax the requirement from a unimodular triangulation to a *unimodular cover* there is indeed a constant c_d depending only on the dimension such that any dilation by a factor of at least c_d has a unimodular cover. This result is from Bruns and Gubeladze but an improvement by von Thaden shows that c_d can be bounded by a polynomial in d.

THEOREM 4.4 ([**26, 136**]). *For each fixed dimension d there is a $c_d \in O(d^6)$ such that $c \cdot P$ has a unimodular cover for every $c \geq c_d$.*

The second half of this section is dedicated to a proof of Theorem 4.1. Our proof differs from the original one (and the reworking of it in [**25**]) in two ways. First,

[1] We call this the Knudsen-Mumford-Waterman Theorem, or KMW, since Knudsen says: "One of the key steps is due to Alan Waterman. The rest is a truly joint effort by Mumford and me" [**77**, p. 109].

a more careful application of the elementary reduction step enables us to avoid using "local lattices" [**25**] or "rational structures" [**77**] and yields a cleaner proof of regularity.[2] Secondly, we set up a book-keeping method that allows us to obtain the bound stated in Theorem 4.5 below. Determining a bound based upon the original proof would be difficult and involve a tower of exponentials whose minimal length would be the maximum volume among the simplices in the triangulation of P being used for the construction (see Remark 4.20). In our proof the bound is "merely" double exponential.

THEOREM 4.5 (Effective KMW Theorem). *If P is a lattice polytope of dimension d and (lattice) volume* $\mathrm{vol}(P)$, *then the dilation*

$$(d+1)!^{\mathrm{vol}(P)!(d+1)^{(d+1)^2 \mathrm{vol}(P)}} P$$

has a regular unimodular triangulation.

More precisely, if P has a triangulation $\mathcal{T}$ into N d-simplices, of volumes $V_1, \dots, V_N$, then the dilation

$$(d+1)!^{\sum_{i=1}^{N} V_i!\left((d+1)!^{d+1}\right)^{V_i-1}} \mathcal{T}$$

has a regular unimodular refinement.

Before going into the KMW Theorem we deal with two questions that are, in a sense, special cases of it. In Section 4.1 we review what is known about KMW-numbers of three-dimensional lattice polytopes, summing up results from [**71**] and [**116**]. In Section 4.2 we give two proofs that if a polytope P has a unimodular triangulation every dilation cP ($c \in \mathbb{N}$) of it has one, and that regularity and flagness of the triangulation can also be preserved in the process. This is equivalent to saying (as noted above) that the set of KMW-numbers of any polytope P is closed under taking products.

The *canonical triangulations* of ordered simplices introduced in Section 4.2 are also instrumental for the proof of Theorem 4.5, which occupies Sections 4.3–4.5. Section 4.3 deals with the case of a lattice d-simplex P and shows that cP can be triangulated into simplices of volume *strictly smaller* than P, for some $c \in \{1, \dots, d+1\}$. In Section 4.4 this reduction is applied iteratively to a triangulation of a general polytope P to yield Theorem 4.1. However, the effective version stated as Theorem 4.5 requires a careful analysis of how to control the number (and type) of new simplices obtained in each iterative step. This is done in our final Section 4.5.

4.1. KMW numbers in dimension three

If P is a segment or a lattice polygon, then P has a regular unimodular triangulation, so for the one and two-dimensional cases, P has KMW-number $c(P) = 1$.

For three-dimensional polytopes the following is known:

(1) One and two are not KMW-numbers of every three-polytope. In fact, there are non-unimodular simplices whose second dilation does not have a unimodular triangulation (Ziegler 1997, unpublished).
(2) It is not known whether three and five are KMW-numbers of every three-polytope.

---

[2] Bruns and Gubeladze [**25**] omit regularity, and the proof of regularity in [**77**] (Theorem 2.22, pp. 147–151) is quite intricate.

(3) Every other integer is a KMW-number of every three-polytope (Kantor and Sarkaria [**71**] for the integer four, Santos and Ziegler [**116**] for every other integer).

These results all emanate from the fact that every full triangulation of P gives a subdivision of cP into dilations of empty simplices. This means that in order to prove that cP can be unimodularly triangulated for every P, we can restrict our attention to the case where P is an empty simplex, as long as we manage to make sure that the triangulations of the individual dilated simplices agree on their common faces. So, let us look at what empty simplices in dimension three look like.

According to White's Theorem [**90, 119, 137**], every empty tetrahedron in $\mathbb{R}^3$ has width one with respect to some lattice direction (not necessarily with respect to a facet normal). That is, it fits between two adjacent lattice hyperplanes. Consequently, it is either unimodular (if it has three vertices in one of the two lattice hyperplanes) or lattice equivalent to

$$S(p,q) := \operatorname{conv}\begin{bmatrix} 0 & 0 & 1 & p \\ 0 & 0 & 0 & q \\ 0 & 1 & 0 & 1 \end{bmatrix}$$

for some integers $0 \le p \le q$ with $gcd(p,q) = 1$. Observe that $S(p,q) \sim S(q-p,q) \sim S(p^{-1},q) \sim S(q-p^{-1},q)$, where inverses are taken modulo q. It can be shown that these are the only cases that give equivalent simplices. The parameter q is the volume of $S(p,q)$, while p carries information of more arithmetic nature.

In the c-th dilation $cS(p,q)$, all the lattice simplices lie in either the bottom edge, the top edge, or the $c-1$ intermediate horizontal lattice planes. The latter can be tiled by lattice translations of the parallelogram $C(p,q) := 2S(p,q) \cap \{z = 1\} = \operatorname{conv}\{(0,0,1),(1,0,1),(p,q,1),(p+1,q,1)\}$. So, in order to understand the structure of lattice points in $cS(p,q)$ we only need to look at those in $C(p,q)$. These are its four vertices plus $q-1$ interior points with the following properties: (a) there is exactly one of them in each of the $c-1$ lattice lines of direction $(1,0,0)$ and intersecting $C(p,q)$ and (b) there is exactly one in each of the $c-1$ lattice lines of direction $(0,p,q)$. This induces two distinct orderings of these interior $q-1$ points which we call the Y-order and the X-order.

All the results about unimodular triangulations of a dilated tetrahedron $S(p,q)$ use, in one way or another, the following decomposition of $cS(p,q)$. We only sketch the descriptions, relying on Figure 4.1 for some of the meaning (more details can be found in [**116**]).

(a) First slice $cS(p,q)$ into c layers by the $c-1$ horizontal planes at integer heights (Figure 4.1.a).

(b) Then divide each layer into a number of triangular prisms in one of two ways: the k-th layer can be divided into either $2k-1$ prisms with axis in the X-direction or $2(c-k)-1$ prisms with axis in the Y-direction (Figure 4.1.b). These prisms have a rectangular facet in one of the horizontal lattice planes and a single edge in the other. We call this edge the *tip* of the prism.

(c) Each such prism can be unimodularly triangulated using joins of the primitive segments in the tip and monotone lattice paths of length q in the opposite facet (Figure 4.1.c). These paths have to be chosen compatible to one another (they can all be the same one) and apart of these joins the triangulation of the prism only contains pyramids, with vertices at the tip, over unimodular triangles in the opposite facet.

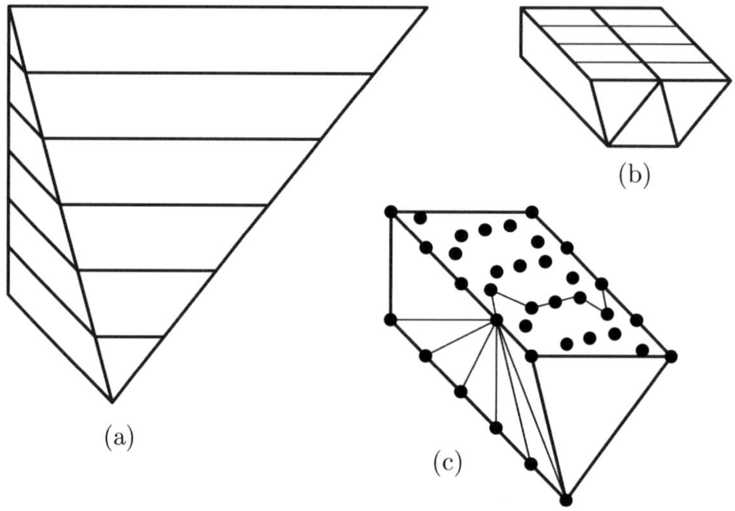

FIGURE 4.1. The three steps in the decomposition of a dilated three-dimensional empty simplex.

With this technique we see that:

COROLLARY 4.6 ([**116**, Cor. 4.2]). *For every lattice three-polytope P and every dilation factor $c \geq 2$, cP can be dissected into unimodular simplices with disjoint interiors.*

PROOF. Consider any full triangulation $\mathcal{T}$ of P, and its dilation $c\mathcal{T}$ into dilated empty simplices. Divide every dilated simplex in layers, then prisms, then simplices, as explained above. □

So, the only thing to take care of is how to make all these dissections agree on the common boundary facets of adjacent dilated simplices, adjacent layers within a dilated simplex, and adjacent prisms within a layer. One way to try to address this issue is to triangulate each dilated simplex $c\Delta$ so that its boundary receives the *standard* triangulation, the one obtained slicing the boundary by all the lattice lines parallel to edges of $c\Delta$ (see Figure 4.2). With this in mind, the following statements on dilated simplices guarantee we can find such desired triangulations of the adjacent components as described above.

THEOREM 4.7. *If $\Delta(p, q)$ is an empty three-simplex, then:*

(1) $2\Delta(p, q)$ *has a unimodular triangulation if and only if* $p = \pm 1 \pmod{q}$. *In this case, the triangulation can be made unimodular and with standard boundary* [**116**, Cor. 3.5, Prop. 3.6]
(2) *For every $c \geq 4$, $c\Delta$ has a unimodular triangulation* ([**116**, Cor. 4.5]).
(3) *For every c that can be written as a sum of composite numbers (that is, every $c \geq 4$, other than $c = 5, 7, 11$), $c\Delta$ has a unimodular triangulation with standard boundary* ([**71**] *for $c = 4$,* [**116**, Thm. 5.1] *for the rest).*
(4) *For every $c \geq 7$, $c\Delta$ has a unimodular triangulation in which the boundary is triangulated with the "quasi-standard" triangulation obtained from the standard one by flipping the diagonals incident to edges of $c\Delta$ and at lattice distance three from the vertices (see Figure 4.2* [**116**]*).*

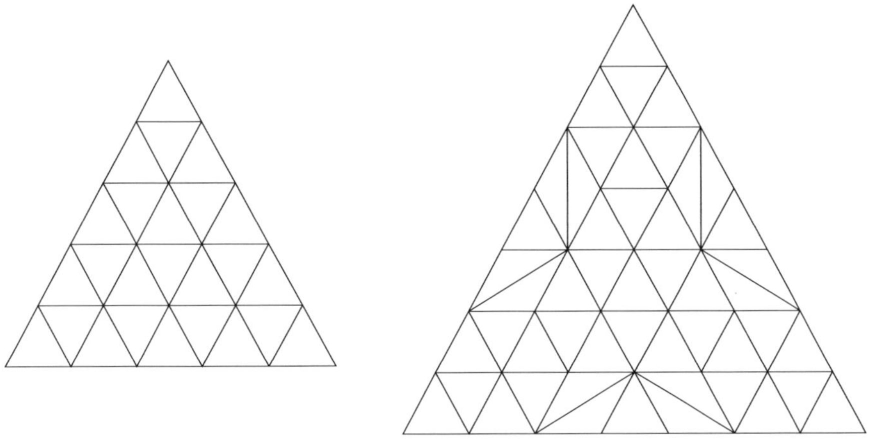

FIGURE 4.2. The standard triangulation of a dilated triangle (left) and the quasi-standard one used for the case $c = 7$.

Hibi, Higashitani, and Yoshida [**65**] generalize part (1) of Theorem 4.7 as follows: it was shown by Batyrev and Hofscheier in [**12**] that empty $(2k-1)$-simplices with unimodular facets are Cayley sums of k unit segments if and only if their k-th dilation does not have interior lattice points. Hibi, Higashitani and Yoshida prove that, among those, the k-th dilation has a unimodular triangulation if and only if they are of the following form:

$$\mathrm{conv}\{O, e_1, \ldots, e_{2k-2}, v\},$$

with $v = e_1 + \cdots + e_{k-1} - e_k - \cdots - e_{2k-2} + qe_{2k-1}$ for some $q \in \mathbb{N}$.

4.2. Canonical triangulation of a dilated simplex

In this section we show that given a unimodular triangulation $\mathcal{T}$ of a polytope P another unimodular triangulation of cP can be derived, preserving regularity (Theorem 4.8) and flagness (Corollary 4.12). The first result can by now be considered "folklore" (it is sometimes attributed to Santos, 1996, unpublished). The second result is new.

The proofs of both theorems start by noting that dilating every simplex in $\mathcal{T}$ yields a triangulation $c\mathcal{T}$ of cP into dilations of unimodular simplices, and then use the canonical subdivisions of dilated unimodular simplices, as discussed in section 2.1.3. That is, the dilated standard simplex $c\Delta^d$ is realized as $\{\boldsymbol{x} \in \mathbb{R}^{d+1}_{\geq 0} : \sum x_i = c\}$, and sliced along the hyperplanes parallel to the facets: $x_i = k$ for $i = 1, \ldots, d+1$ and $k = 1, \ldots c-1$. The cells of this subdivision are hypersimplices. The canonical weights $\boldsymbol{\omega_m} := \sum m_i^2$ show that this subdivision is regular, and the weights restrict to faces as well.

The naturality of this subdivision and its weights allows one to patch simplices together. That is, if we subdivide each of the cells of $c\mathcal{T}$ canonically, they agree along their boundaries, and we get a subdivision of P which is regular if $\mathcal{T}$ was regular.

THEOREM 4.8. *If P has a (regular) unimodular triangulation $\mathcal{T}$ then, for every positive integer c, its dilation cP has one too.*

PROOF. As in Theorem 2.4, all cells (hypersimplices) of the canonical subdivision of $c\mathcal{T}$ are compressed. Hence, pulling all the lattice points of cP in any order yields a unimodular triangulation $\mathcal{T}'$. This triangulation is a regular refinement. Hence, regularity of $\mathcal{T}$ implies regularity of $\mathcal{T}'$ (cf. [**44**, Lemmas Lemma 2.3.16 and 4.3.12], or see more details in the proof of Corollary 4.12). □

To guarantee flagness, instead of using pulling refinements we dice with respect to a larger set of hyperplanes, namely those based on the root system of type **A**. Remember that if P is a lattice polytope of type **A** then dicing P according to all lattice hyperplanes with normals in the root system gives a quadratic triangulation of P (Theorem 3.3).

To apply this to an arbitrary simplex Δ we first need to show a way to map a d-simplex to a simplex of type **A**. This is canonical, once an ordering of Δ's vertices is prescribed. Thus we introduce the concept of an ordered simplex, and the canonical triangulation of its dilation. These concepts will be crucial in the proof of the KMW Theorem (and were used in the previous proofs of it, see [**25**, Section 3.A]).

DEFINITION 4.9. An *ordered simplex* is a simplex, $\Delta = \mathrm{conv}\{\boldsymbol{a}_0, \dots, \boldsymbol{a}_d\}$ that comes with an ordering of its vertices, $\boldsymbol{a}_0, \dots, \boldsymbol{a}_d$. Let Δ be an ordered d-simplex and consider the affine isomorphism sending Δ to the standard simplex $\Delta^d_{\mathbf{A}}$ of type **A** via

$$\boldsymbol{a}_i \mapsto \sum_{j=1}^{i+1} \boldsymbol{e}_j \in \mathbb{R}^{d+1}.$$

The **A**-*canonical triangulation* of $c\Delta$ is the pull back under this map of the type **A** dicing triangulation of $c\Delta^d_{\mathbf{A}}$.

Observe that since the **A**-dicing is unimodular, all the simplices in the canonical triangulation of $c\Delta$ have the same volume as Δ.

The following lemma makes what the canonical triangulation is more explicit (as a dicing triangulation) by expressing it in barycentric coordinates with respect to Δ. Recall that the barycentric coordinates of a point $\boldsymbol{x} \in \Delta$ with respect to the vertices $\boldsymbol{a}_0, \dots, \boldsymbol{a}_d$ of a simplex Δ are the unique vector $(x_0, \dots, x_d)$ with $\sum x_i = 1$ such that

$$\boldsymbol{x} = \sum_{i=0}^{d} x_i \boldsymbol{a}_i.$$

In particular, for each subset $I \subseteq \{0, \dots, d\}$ the hyperplanes defined by setting $\sum_{i \in I} x_i$ equal to a constant are parallel to the following two complementary faces of Δ:

$$\mathrm{conv}\{\boldsymbol{a}_i : i \in I\}, \qquad \text{and} \qquad \mathrm{conv}\{\boldsymbol{a}_i : i \notin I\}.$$

LEMMA 4.10. *Let $\Delta = \mathrm{conv}\{\boldsymbol{v}_0, \dots, \boldsymbol{v}_d\}$ be an ordered simplex and c a positive integer. Then the canonical triangulation of Δ is the dicing triangulation with respect to the families of lattice hyperplanes defined by making any sum $x_i + \cdots + x_j$ of consecutive barycentric coordinates constant.*

PROOF. Since an affine isomorphism from one simplex to another preserves barycentric coordinates, there is no loss of generality in assuming that Δ is the standard simplex $\Delta^d_{\mathbf{A}}$. We only need to show that the hyperplanes orthogonal to the roots are indeed defined by making a sum of consecutive barycentric coordinates constant. For this, observe that if $(x_0, \dots, x_d)$ are the barycentric coordinates of $\boldsymbol{x}$ with respect to $\Delta^d_{\mathbf{A}}$ then

$$\boldsymbol{x} = \sum_{j=1}^{d} x_j \left(\sum_{i=1}^{j+1} \boldsymbol{e}_i \right) = \sum_{i=1}^{d+1} \boldsymbol{e}_i \left(\sum_{j=i-1}^{d} x_j \right),$$

so that

$$\langle \boldsymbol{x}, \boldsymbol{e}_j - \boldsymbol{e}_i \rangle = \sum_{k=i-1}^{j-2} x_k, \qquad \forall 1 \le i < j \le d+1.$$

□

With this we can conclude the following.

THEOREM 4.11. *Let $\mathcal{T}$ be a lattice triangulation of a polytope P, let c be a positive integer, and let $c\mathcal{T}$ denote the dilation of $\mathcal{T}$, that is, the following triangulation of cP:*

$$c\mathcal{T} := \{c\Delta : \Delta \in \mathcal{T}\}.$$

Consider a total order in the vertices of $\mathcal{T}$, so that every $\Delta \in \mathcal{T}$ is regarded as an ordered simplex. Call $\mathcal{T}'$ the triangulation of cP obtained refining each $c\Delta$ to its canonical triangulation. Then:

1. *$\mathcal{T}'$ is indeed a triangulation. That is, common faces of different simplices in $c\mathcal{T}$ get the same refinement.*
2. *If $\mathcal{T}$ his unimodular, then $\mathcal{T}'$ is unimodular.*
3. *If $\mathcal{T}$ is regular, then $\mathcal{T}'$ is regular.*
4. *If $\mathcal{T}$ is flag, then $\mathcal{T}'$ is flag.*

PROOF. The canonical triangulations agree on common faces of simplices of $c\mathcal{T}$ since they only depend on the ordering of the vertices in those faces. Therefore, this procedure indeed gives a lattice triangulation $\mathcal{T}'$ of cP. Since the canonical triangulation preserves volumes, it preserves unimodularity. We will show that regularity and flagness are also preserved.

For regularity we use the idea of regular refinements [**44**, Lemma 2.3.16]. Assume that $\mathcal{T}_0$ is a regular subdivision of a point configuration $\mathcal{A}$, given by a weight vector $\boldsymbol{\omega}_0$. If each cell of $\mathcal{T}_0$ is refined using a common weight vector $\boldsymbol{\omega} \in \mathbb{R}^{\mathcal{A}}$, then the refinements of all cells agree on their intersections, so that we get a subdivision of $\mathcal{A}$. Moreover, this subdivision is regular, and could have been obtained directly using the weight vector $\boldsymbol{\omega}_0 + \epsilon\boldsymbol{\omega}$, for a sufficiently small $\epsilon > 0$.

We apply this procedure $\binom{n+1}{2}$ times to the subdivision $c\mathcal{T}$, where n is the number of vertices of $\mathcal{T}$, as follows. Let $\boldsymbol{a}_1, \dots, \boldsymbol{a}_n$ be the ordered sequence of vertices of $\mathcal{T}$. For each interval $[i,j] \subseteq [1,n]$ (including the case $i = j$), let $\phi_{i,j}$ be the function that is zero on $c\boldsymbol{a}_k$ if $k \in [i,j]$, one on $c\boldsymbol{a}_k$ if $k \notin [i,j]$, and linear on each simplex of $c\mathcal{T}$. Consider the weight function $\boldsymbol{\omega}_{i,j} \in \mathbb{R}^{cP\cap\mathbb{Z}^d}$ that restricts $\phi^2_{i,j}$ to all lattice points in cP. Clearly, the regular refinement of $c\mathcal{T}$ via this weight function dices each simplex $c(\operatorname{conv}\{\boldsymbol{a}_i : i \in S\})$ by the lattice hyperplanes parallel to $\operatorname{conv}\{\boldsymbol{a}_i : i \in S \cap [i,j]\}$ and $\operatorname{conv}\{\boldsymbol{a}_i : i \in S \setminus [i,j]\}$. In each simplex of $c\mathcal{T}$ this is one of the hyperplanes we want to use for the canonical triangulation of that simplex. Performing these regular refinements on $c\mathcal{T}$ for each such weight function $\boldsymbol{\omega}_{i,j}$, in any order, gives a dicing of each simplex by more and more families of hyperplanes, eventually all hyperplanes in Lemma 4.10. Hence, the resulting triangulation is $\mathcal{T}'$, proving that $\mathcal{T}'$ is regular.

Flagness of $\mathcal{T}'$ is based solely on the fact that $\mathcal{T}$ is flag and that the way $\mathcal{T}'$'s refines each simplex of $c\mathcal{T}$ is also flag (by Theorem 3.3). Indeed, let $N \subset cP \cap \mathbb{Z}^d$ and suppose that every pair of points in N form an edge in $\mathcal{T}'$. To each point $\boldsymbol{n} \in N$ associate its carrier simplex $cF(\boldsymbol{n}) \in c\mathcal{T}$, and we call S the union of all vertices of all the carriers $F(\boldsymbol{n})$. Observe that for $\boldsymbol{n}_1, \boldsymbol{n}_2 \in N$ we have that $cF(\boldsymbol{n}_1) \cup cF(\boldsymbol{n}_2)$ equals the minimal face of $c\mathcal{T}$ containing the edge $\{\boldsymbol{n}_1, \boldsymbol{n}_2\} \in \mathcal{T}'$ (here we are using the fact that $\mathcal{T}'$ refines $c\mathcal{T}$). In particular, every two points in S form an edge of $\mathcal{T}$ because either they lie in the same $F(\boldsymbol{n})$, which is a simplex, or they lie in $F(\boldsymbol{n}_1)$ and $F(\boldsymbol{n}_2)$ for two points of N. Since $\mathcal{T}$ is flag, S is a face of $\mathcal{T}$, which implies that N is contained in the convex hull of the face $cS \in c\mathcal{T}$. Since $\mathcal{T}'$ refines cS in a flag manner and N is a clique in $\mathcal{T}'$, N is a face in $\mathcal{T}'$. $\square$

COROLLARY 4.12. *If P has a quadratic triangulation $\mathcal{T}$, then so does its dilation cP for every positive integer c.*

With the canonical triangulations at our disposal, we can now discuss the example advertised at the beginning of the section. This example is inspired by [**91**] and appears in [**40**].

EXAMPLE 4.13. Let h and k be positive integers and set $d := hk - 1$. In $\mathbb{R}^{d+1}$ consider the vector $v := \frac{1}{k}(1, \dots, 1)$, the lattice $\Lambda := \mathbb{Z}^{d+1} + \mathbb{Z}v$ and the d-polytope $\Delta := \operatorname{conv}\{e_0, \dots, e_d\}$. Then $h\Delta$ has a quadratic triangulation joining v to the canonical triangulation of $h\partial\Delta$.

We claim that certain higher multiples of Δ are not normal and thus do not admit unimodular triangulations. To see this, consider the homomorphism $\phi\colon \Lambda \to \mathbb{Z}/k$ given by $x \mapsto \overline{kx_0}$. If a and b are positive integers with $b < h$, then the Λ-points in $(ah+b)\Delta$ map to $\{\bar{0}, \bar{1}, \dots, \bar{a}\}$. Therefore, we can only obtain $\{\bar{0}, \bar{1}, \dots, \overline{2a}\}$ from the sum of two such points. On the other hand, if $2b \geq h$, then $2(ah+b)\Delta$ contains a Λ-point with image $\overline{2a+1}$, witnessing non-normality if $2a+1 < k$. The parameters $k = 4$, $h = 2$ yield a seven-simplex so that 2Δ has a quadratic triangulation but 3Δ is not normal.

EXAMPLE 4.14. Concerning the IDP property, Lasoń and Michałek [**80**, Sect. 3.1–3.3] show the following: let $T(k)$ be the simplex in $\mathbb{R}^{2k-1}$ obtained as the convex hull of the origin, the first $2k-2$ coordinate unit vectors $e_1, \dots, e_{2k-2}$, and the point $v := e_1 + \dots + e_{k-1} - e_k - \dots - e_{2k-2} + (k+2)e_{2k-1}$ in $\mathbb{R}^{2k-1}$. (Equivalently, $T(k)$ is obtained from a unimodular (k,k) circuit in $\mathbb{R}^{2k-2}$ by lifting one point to height $k+2$ and the rest to height 0). Let $P(k)$ be the Minkowski sum of $T(k)$ and the segment $[O, e_{2k-1}]$. Then, $cP(k)$ is IDP if and only if $c \geq k$ or c does not divide k. For example, $2P(25)$ and $3P(25)$ are IDP, but $5P(25)$ is not, for the 49-dimensional polytope $P(25)$.

4.3. Reducing the volume of simplices in the dilation

The canonical triangulation divides $c\Delta$ into simplices of the same volume as Δ. We now want to improve this, and triangulate $c\Delta$ into simplices of volume strictly smaller than Δ (assuming Δ is not unimodular). This cannot be done for every c, but we prove here that it can always be done for some $c \leq d$ and all its multiples.

For this we introduce the concept of box points.

We start with a seemingly unrelated lemma about the Cayley sum of two dilated polytopes of type **A**. Remember that if P_1 and P_2 are lattice polytopes in a lattice $\Lambda \subset \mathbb{R}^d$ then the *Cayley sum* of P_1 and P_2 is the lattice polytope $\operatorname{conv}(P_1 \times \{0\} \cup P_2 \times \{1\})$ in the lattice $\Lambda \times \mathbb{Z} \subset \mathbb{R}^{d+1}$.

As usual, the canonical triangulation of a polytope of type **A** is the one obtained by slicing by all the lattice hyperplanes normal to the roots. Remember that this triangulation is regular and unimodular (Theorem 3.3).

LEMMA 4.15. *If P_1 and P_2 are lattice polytopes of type* **A**, *with canonical triangulations $\mathcal{T}_1$ and $\mathcal{T}_2$, then:*

(1) *There is a regular triangulation of the Cayley sum* $\operatorname{conv}(P_1 \times \{0\} \cup P_2 \times \{1\})$ *that restricts to $\mathcal{T}_1$ and $\mathcal{T}_2$ on P_1 and P_2.*
(2) *Any such triangulation (regular or not) is unimodular.*

PROOF. The canonical triangulation of a type-**A** polytope is regular, so let $\boldsymbol{\omega}_1$ and $\boldsymbol{\omega}_2$ be weight vectors producing $\mathcal{T}_1$ and $\mathcal{T}_2$ in P_1 and P_2, respectively. Since $P_1 \times \{0\}$ and $P_2 \times \{1\}$ are faces of the Cayley sum, using the same weight vectors (that is, using $\boldsymbol{\omega}_1$ in $P_1 \times \{0\}$ and $\boldsymbol{\omega}_2$ in $P_2 \times \{1\}$) yields a regular subdivision that restricts to $\mathcal{T}_1$ and $\mathcal{T}_2$ on P_1 and P_2. If that subdivision is not a triangulation, we do pulling refinements, which preserve regularity.

For part (2), observe that every cell in such a triangulation $\mathcal{T}$ will be a join of a face F_1 of $\mathcal{T}_1$ and a face F_2 of $\mathcal{T}_2$. It is convenient to think of the root system $\mathbf{A}_d$ as consisting of the standard basis vectors $\pm\boldsymbol{e}_i$ plus the differences $\pm(\boldsymbol{e}_i - \boldsymbol{e}_j)$. This is done by forgetting the coordinate $d+1$ in the usual definition of $\mathbf{A}_d$ as embedded in $\mathbb{R}^{d+1}$. In this description, the $\mathbf{A}_d$ dicing of $\mathbb{R}^d$ refines the tiling by translations of the unit cube, triangulating every lattice unit cube by (translations of) the order triangulation of $[0,1]^d$.

In particular, if k and l are the dimensions of F_1 and F_2 there is no loss of generality in assuming that

$$F_1 = \operatorname{conv}\{\boldsymbol{a}_0, \boldsymbol{a}_1 \dots, \boldsymbol{a}_k\} \times \{0\}, \qquad F_2 = \operatorname{conv}\{\boldsymbol{b}_0, \boldsymbol{b}_1 \dots, \boldsymbol{b}_l\} \times \{1\},$$

where the $\boldsymbol{a}_i$'s and the $\boldsymbol{b}_j$'s are sequences of $0/1$ vectors in $\mathbb{R}^d$ with increasing supports. Moreover, by subtracting $\boldsymbol{a}_0$ from every $\boldsymbol{a}_i$ and $\boldsymbol{b}_0$ from every $\boldsymbol{b}_j$ we can assume $\boldsymbol{a}_0 = \boldsymbol{b}_0 = 0$. Observe that the $\boldsymbol{a}_i$ and $\boldsymbol{b}_i$ are in transversal spaces. Then, the volume of the join of F_1 and F_2 equals the volume of $F := \operatorname{conv}\{0 = \boldsymbol{a}_0 = \boldsymbol{b}_0, \boldsymbol{a}_1 \dots, \boldsymbol{a}_k, \boldsymbol{b}_1 \dots, \boldsymbol{b}_l\} \subset \mathbb{R}^d$. We are going to show that this volume is always one (or zero). Without loss of generality we assume F to be full-dimensional. If it is not, we extend the sequence of $\boldsymbol{a}_i$'s until it is.

We use induction on the dimension d. In particular, we assume that no $\boldsymbol{a}_i$ and $\boldsymbol{a}_{i+1}$ differ in a single coordinate, because otherwise we can project along that coordinate and get the result by inductive hypothesis. The same happens if some $\boldsymbol{b}_j$ and $\boldsymbol{b}_{j+1}$ differ by a single coordinate. That implies that both k and l are at most $d/2$, hence they are both equal to $d/2$ because $k + l = d$. But then we must have $\boldsymbol{a}_k = \boldsymbol{b}_l = (1, \dots, 1)$, which implies that the volume of F is zero (except, of course, in the base case for the induction, which is $k = l = 0$ and produces volume one). □

It is worth mentioning that Lemma 4.15 is not true for the Cayley sum of three or more type-**A** polytopes, even in dimension three. Indeed, the edges parallel to the vectors $(1,1,0)$, $(1,0,1)$ and $(0,1,1)$ are type-**A** polytopes and their Cayley sum, which equals their join, is a non-unimodular simplex.

Our method for refining dilations will be based on combining Lemma 4.15 with the concept of an **A**-canonical triangulation of a dilated ordered simplex (Definition 4.9). Let us explain how.

Let $\Delta = \operatorname{conv}(\boldsymbol{b}_0, \dots, \boldsymbol{b}_d)$ be a non-unimodular lattice simplex, of volume V with respect to a lattice Λ. Let L_Δ be the lattice spanned by the vertices of Δ. We take $\vec{L}_\Delta := L_\Delta$ to denote the linear lattice parallel to L_Δ, so $\Lambda : \vec{L}_\Delta = V$.[3] We use barycentric coordinates with respect to Δ, and dilated barycentric coordinates for a dilation $c\Delta$. In the latter, points in $c\Delta$ are written as convex combinations of $c\boldsymbol{v}_0, \dots, c\boldsymbol{v}_d$.

A *box point* for Δ is simply an element of the quotient $\Lambda/\vec{L}_\Delta$, where $\vec{L}_\Delta$ is the linear lattice parallel to L_Δ. We normally represent a box point in the Δ-barycentric coordinates of any of its representatives, as $\boldsymbol{m} = (m_0, \dots, m_d)$. Quotienting by $\vec{L}_\Delta$ then means that we are interested only in $(\{m_0\}, \dots, \{m_d\})$, where $\{x\} := x - \lfloor x \rfloor$ denotes the fractional part of a real number x. The *height* $h(\boldsymbol{m})$ of a box point $\boldsymbol{m}$ is the number $\sum_i \{m_i\}$. It is an integer between one and $V - 1$ and coincides with the smallest integer h for which $h\Delta$ contains a point of $\boldsymbol{m} + \Lambda$.

These concepts extend to lower dimensional faces. For a face F L_F denotes the lattice spanned by the vertices of F and Λ_F is the intersection of Λ with the linear space parallel to F. A box point for F is then an element of $\Lambda_F/\vec{L}_F$, and is represented by an $\boldsymbol{m}$ in barycentric coordinates with respect to F. Observe that if F is a face of Δ then there is a natural inclusion $\Lambda_F/\vec{L}_F \le \Lambda/(\vec{L}_\Delta)$ so that a box point for F is also a box point for Δ. Coordinates for $\boldsymbol{m}$ are extended from F to Δ by putting zeroes in the coordinates of the vertices in $\Delta \setminus F$. Conversely, a box point of Δ is a box point for F if and only if those particular coordinates are zero.

We call *carrier* of a box point $\boldsymbol{m}$ of Δ the (unique) minimal face F of Δ for which $\boldsymbol{m}$ is a box point. A box point of Δ is *non-degenerate* if its carrier is Δ itself. That is, if none of its Δ-barycentric coordinates is integral.

Given a box point $\boldsymbol{m}$ of a simplex Δ and a certain positive integer c, we aim to triangulate $c\Delta$ into simplices of volume strictly smaller than that of Δ, using only the points of $L_\Delta \cup (\boldsymbol{m} + L_\Delta)$. Of course, this is impossible if $c < h(\boldsymbol{m})$. Although it can be done for every $c \ge h(\boldsymbol{m})$

We offer two versions of this procedure. The first works for every $c \ge c_0$ but is less symmetric and complicates the proof of regularity. The second one is somehow simpler but requires c to be a multiple of $(d+1)!$ (or, rather, a multiple of the number of non-zero coordinates in $\boldsymbol{m}$. However, since we will later apply the procedure to several box points at the same time, having c be a multiple of $(d+1)!$ is convenient).

The structure of $c\Delta \cap (\boldsymbol{m} + L_\Delta)$ is straightforward: it is a translation of the points of L_Δ in $(c - c_0)\Delta$ (see Figure 4.3).

Assuming Δ is an ordered simplex allows us to speak of the canonical triangulation of $c\Delta$, and of cF for every face F of Δ.

Let F be a face in a triangulation $\mathcal{T}$ and let $\boldsymbol{m}$ be a box point of F. We say that $\boldsymbol{m}$ is *non-degenerate* for F if it is not a box point of a proper face of F. Equivalently, $\boldsymbol{m}$ is non-degenerate if all the F-coordinates of $\boldsymbol{m}$ are fractional. Then, the full-dimensional simplices of $\mathcal{T}$ for which $\boldsymbol{m}$ is a box point are precisely

[3]To simplify the exposition, in the rest of the paper we assume (w. l. o. g.) that Λ contains the origin, so that all the dilations $c\Delta$ are lattice polytopes with respect to the same lattice Λ. We do not require, however, the affine lattice L_Δ to contain the origin.

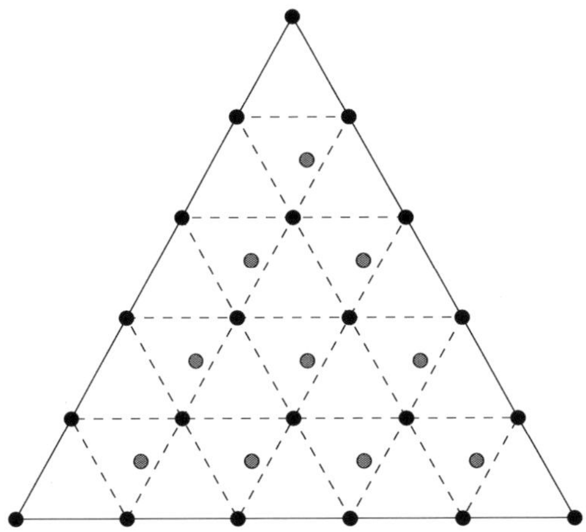

FIGURE 4.3. The points in $c\Delta \cap (\boldsymbol{m} + L_\Delta)$. Here $c = 5$ and $c_0 = 2$

the star of F. Recall that the (closed) star of a face F in a simplicial complex $\mathcal{T}$ is the set $\overline{\operatorname{star}}(F;\mathcal{T})$ of all simplices of $\mathcal{T}$ that contain F, together with all their faces. Equivalently:

$$\overline{\operatorname{star}}(F;\mathcal{T}) := \{F' \in \mathcal{T} : F \cup F' \in \mathcal{T}\}.$$

By $\partial \operatorname{star}(F;\mathcal{T})$ we mean the faces of $\overline{\operatorname{star}}(F;\mathcal{T})$ that do not contain F. If $\mathcal{T}$ is a triangulation of a polytope and F is not contained in the boundary of P then $\partial \operatorname{star}(F;\mathcal{T})$ coincides with the topological boundary of $\overline{\operatorname{star}}(F;\mathcal{T})$.

As before, a global ordering is assumed on the vertices of $\mathcal{T}$, so that we can speak of the canonical triangulation of each dilated face in $c\mathcal{T}$.

LEMMA 4.16. *If $\mathcal{T}$ is a lattice triangulation on an ordered set of vertices lying in a lattice Λ, and $F = \{\boldsymbol{v}_0, \dots, \boldsymbol{v}_k\}$ is a non-unimodular face with a non-degenerate box point $\boldsymbol{m} = (m_0, \dots, m_k) \in L_F \setminus \Lambda_F$, then for every integer $c \in (k+1)\mathbb{N}$, $c \cdot \overline{\operatorname{star}}(F;\mathcal{T})$ has a refinement $\mathcal{T}_{\boldsymbol{m}}$ such that:*

1. *The volume of every full-dimensional simplex Δ' in $\mathcal{T}_{\boldsymbol{m}}$ is strictly less than the volume of simplex Δ for which $\Delta' \subset c\Delta$.*
2. *$\mathcal{T}_{\boldsymbol{m}}$ induces the canonical triangulation on the boundary $c \cdot \partial \operatorname{star}(F;\mathcal{T})$.*
3. *$\mathcal{T}_{\boldsymbol{m}}$ is a regular refinement of $\mathcal{T}$, so if $\mathcal{T}$ is regular then $\mathcal{T}_{\boldsymbol{m}}$ is regular. In particular, any choice of weights inducing the canonical triangulation on $\Lambda \cap c \cdot \partial \operatorname{star}(F;\mathcal{T})$ has an extension to $\Lambda \cap c \cdot \overline{\operatorname{star}}(F;\mathcal{T})$ that induces $\mathcal{T}_{\boldsymbol{m}}$ as a regular refinement of $\mathcal{T}$.*

PROOF. The idea is to first subdivide each $c\Delta$, $\Delta \in \mathcal{T}$ by concentric copies of smaller and smaller dilations of Δ and then use Lemma 4.15 to refine those subdivisions. But let us explain it in a way that demonstrates the regularity properties.

We start by describing the structure of lattice points in $c \cdot \overline{\operatorname{star}}(F;\mathcal{T})$. (Note: these are not all the lattice points; we are not claiming our triangulation $\mathcal{T}_{\boldsymbol{m}}$ to be full.) Observe that requiring c to be a multiple of k implies that the barycenter of cF is in L_F. Around it, we have concentric copies $k\partial \operatorname{star}(F;\mathcal{T})$, $2k\partial \operatorname{star}(F;\mathcal{T})$, $\dots$, $c \cdot \partial \operatorname{star}(F;\mathcal{T})$, which contain all the lattice points of each $c\Delta$ that lie in L_Δ, for each $\Delta \in \overline{\operatorname{star}}(F;\mathcal{T})$. Between each pair of consecutive copies of multiples of $k\partial \operatorname{star}(F;\mathcal{T})$, we have translated copies of $(k - c_0)\partial \operatorname{star}(F;\mathcal{T})$,

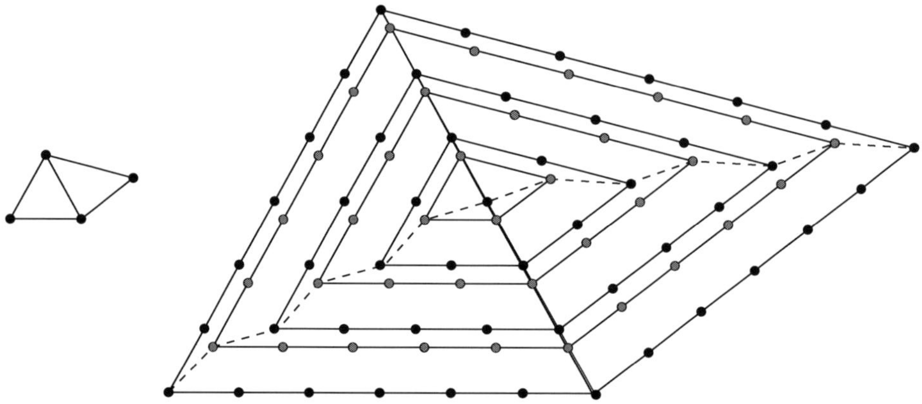

FIGURE 4.4. $\overline{\text{star}}(F;\mathcal{T})$ (left) and the subdivision of $c\cdot\overline{\text{star}}(F;\mathcal{T})$ into Cayley polytopes via concentric dilated copies of $\partial\,\text{star}(F;\mathcal{T})$ (right)

$(2k-c_0)\partial\,\text{star}(F;\mathcal{T})$, ..., $(c-c_0)\partial\,\text{star}(F;\mathcal{T})$, and those complexes contain the lattice points in the distinct $\boldsymbol{m}+L_\Delta$. The latter are not concentric to the barycenter, but concentric to one another and displaced by the vector $\boldsymbol{m}$. For each two consecutive dilated copies of $\partial\,\text{star}(F;\mathcal{T})$, corresponding faces form a Cayley polytope and these Cayley polytopes form a polyhedral subdivision of $c\cdot\overline{\text{star}}(F;\mathcal{T})$ (the innermost copy $(c-c_0)\overline{\text{star}}(F;\mathcal{T})$ is stellarly subdivided from the origin into pyramids, which are nothing but degenerate cases of Cayley polytopes).

We claim that this polyhedral subdivision $\mathcal{S}$ is the regular refinement of $c\cdot\overline{\text{star}}(F;\mathcal{T})$ obtained by the following choice of weights. Choose numbers $\alpha_0=0<\alpha_1<\alpha_2<\cdots<\alpha_{2c/k}$, each much bigger than the previous one, and give height α_i to the points in the i-th concentric copy.

We then refine $\mathcal{S}$ into $\mathcal{T}_{\boldsymbol{m}}$ via any second choice of weights $\boldsymbol{\omega}$. Any such refinement will satisfy the first claimed property. If $\boldsymbol{\omega}$ is chosen to induce the canonical triangulation on $c\cdot\partial\,\text{star}(F;\mathcal{T})$, then $\mathcal{T}_{\boldsymbol{m}}$ also satisfies properties (2) and (3). For (3), observe that our first choice of weights (the α's) were constant on $c\cdot\partial\,\text{star}(F;\mathcal{T})$. Hence, the restriction of $\mathcal{T}_{\boldsymbol{m}}$ to $c\cdot\partial\,\text{star}(F;\mathcal{T})$ is only governed by $\boldsymbol{\omega}$. □

Let us look at how much the number of simplices in a triangulation grows with each iteration of Lemma 4.16. One crude bound is that the number of simplices produced in the refinement of each individual dilated simplex $c\Delta$ is at most the total number of simplices in the refinement.

LEMMA 4.17. *In the triangulations constructed in Lemma* 4.16 *and Corollary* 4.18, *each dilated simplex* $c\Delta$, $\Delta\in\mathcal{T}$, *is refined into at most* $(d+1)c^d$ *full-dimensional simplices.*

PROOF. Observe that the combinatorial type, hence the number of simplices, of $\mathcal{T}_m$ depends only on two parameters (apart of d and c): how many of the $\{m_i\}$'s are not zero (call this number d_0), and the value of $c_0=\sum_i\{m_i\}$. In particular, to compute the number of simplices there is no loss of generality in assuming that all the non-zero $\{m_i\}$'s are equal to one another, hence equal to c_0/d_0. In this case, all the simplices in $\mathcal{T}_{\boldsymbol{m}}$ have volume exactly c_0/d_0 times the volume of Δ, so the number of them needed to fill $c\Delta$ equals

$$\frac{d_0 c^d}{c_0} \leq (d+1)c^d.$$

In the last inequality we use that $d_0 \leq d+1$ and $c_0 \geq 1$. □

4.4. A proof of the KMW Theorem

Lemma 4.16 can be applied simultaneously to several faces $F_1, \ldots, F_N$ each with a non-degenerate box point $\boldsymbol{m}_1, \ldots, \boldsymbol{m}_N$, as long as the stars of the F_i's intersect only on their boundaries. Equivalently, as long as there is not an i, j pair for which $\operatorname{conv}(F_i \cup F_j)$ is a simplex, or as long as the open stars of the F_i's are disjoint. Here the *open star* of a face in a simplicial complex is defined as:

$$\operatorname{star}(F; \mathcal{T}) = \{F' \in \mathcal{T} : F \subset F'\}.$$

(Observe that open stars are not simplicial complexes.)

In the following statement, we assume that c is a multiple of $(d+1)!$ to guarantee that it is a multiple of $\dim(F_i) + 1$ for every F_i, as required in order to apply Lemma 4.16.

COROLLARY 4.18. *Let $\mathcal{T}$ be a lattice triangulation (with an ordering of its vertices). Suppose there is a family $F_1, \ldots, F_N$ of faces of $\mathcal{T}$ with respective non-degenerate box points $\boldsymbol{m}_1, \ldots, \boldsymbol{m}_N$ such that $\operatorname{conv}(F_i \cup F_j) \notin \mathcal{T}$, for every pair $i, j \in [N]$.*

Then, for every integer $c \in (d+1)!\mathbb{N}$, the dilation $c\mathcal{T}$ can be refined into a triangulation $\mathcal{T}'$ with the following properties:

1. *$\mathcal{T}'$ is the canonical refinement of $c\mathcal{T}$ away from $\cup_i \overline{\operatorname{star}}(F_i; \mathcal{T})$. In particular, for every full-dimensional simplex $\Delta' \in \mathcal{T}'$ not in $\cup_i \overline{\operatorname{star}}(F_i; \mathcal{T})$, $L_{\Delta'} = L_\Delta$, where $\Delta \in \mathcal{T}$ is such that $\Delta' \subset c\Delta$.*
2. *For each full-dimensional simplex Δ' of $\mathcal{T}'$ contained in $\cup_i \overline{\operatorname{star}}(F_i; \mathcal{T})$, $\operatorname{vol}(\Delta') < \operatorname{vol}(\Delta)$ for $\Delta \in \mathcal{T}$ is such that $\Delta' \subset c\Delta$.*
3. *$\mathcal{T}'$ is a regular refinement of $\mathcal{T}$.*

PROOF. The condition $\operatorname{conv}(F_i \cup F_j) \notin \mathcal{T}$ is equivalent to saying that the closed stars $\overline{\operatorname{star}}(F_i; \mathcal{T})$ and $\overline{\operatorname{star}}(F_j; \mathcal{T})$ intersect only on their boundaries. So, Lemma 4.16 can be applied simultaneously with respect to all box points and stars, since it gives the canonical triangulation on the boundary.

We know the components match up to form a regular refinement since refining every simplex of $c\mathcal{T}$ to its canonical triangulation is clearly a regular refinement. By part (3) of Lemma 4.16, if we keep the weights that give the canonical triangulations outside the stars of the F_i's (including their boundaries) we can extend that to give our desired refinement inside the stars. □

With this we can conclude the following statement, which implies Theorem 4.1 by induction on V.

THEOREM 4.19. *Let $\mathcal{T}$ be a triangulation of P, and V be the maximal volume among its simplices. If $\Delta_1, \ldots, \Delta_N$ is the full-dimensional simplices of volume V, then applying Corollary 4.18 N times with any $c \in (d+1)!\mathbb{N}$ yields a regular refinement of $c^N\mathcal{T}$ into a triangulation with maximal volume strictly less than V.*

PROOF. Our goal is to apply apply Corollary 4.18 once (at most) for each of the N simplices of volume V. The only difficulty is that with each application the number of simplices of volume V grows. Indeed, for each such simplex Δ not lying in the star of one of the F_i's that we use in a particular step, $c\Delta$ is refined into its canonical triangulation, which consists of a (huge) number of simplices of the same volume V. Our proof takes advantage of the fact that all these simplices form a canonical triangulation, to refine (dilations of) all of them simultaneously into simplices of volume strictly smaller than V.

Suppose that a triangulation $\mathcal{T}$ contains the canonical triangulation $\mathcal{T}_{k\Delta}$ of a dilated simplex $k\Delta$, for some k. Choose a box point $\boldsymbol{m}_0$ of one of the simplices $\Delta_0 \in \mathcal{T}_{k\Delta}$. Since all the simplices of the canonical triangulation define the same lattice, $\boldsymbol{m}_0$ can also be considered a box point in every other simplex of $\mathcal{T}_{k\Delta}$. So, let $F_1, \ldots, F_M$ ($M \leq k^d$) be the list of minimal faces of $\mathcal{T}_{k\Delta}$ for which $\boldsymbol{m}_0$ is a box point. We claim that the stars of the F_i's are disjoint and that they cover $\mathcal{T}_{k\Delta}$. (The second claim is obvious; the first is also not hard, and a proof of it in a more general context is in Lemma 4.25 below.) Hence, we can get rid off all the simplices of $\mathcal{T}_{k\Delta}$ in a single step.

So, our algorithm for refining $c^N\mathcal{T}$ in N steps is: In the first step, we choose a box point $\boldsymbol{m}_1$ for Δ_1, and apply Corollary 4.18 to that box point alone, to get a triangulation $\mathcal{T}_1$ in which all the simplices of volume V belong to the canonical triangulations of $c\Delta_2$, $c\Delta_3, \ldots, c\Delta_N$. After the i-th step, we have a triangulation $\mathcal{T}_i$ in which all the simplices of volume V belong to the canonical triangulations of $c^i\Delta_{i+1}$, $\ldots$, $c^i\Delta_N$. We choose a box point $\boldsymbol{m}_{i+1}$ for one simplex in $c^i\Delta_{i+1}$ and apply Corollary 4.18 simultaneously to the stars of all the minimal faces of $c^i\Delta_{i+1}$ having $\boldsymbol{m}_{i+1}$ as a box point. □

REMARK 4.20. As pointed out by Bruns and Gubeladze [**25**, Remark 3.20], "One could try to give an effective upper bound for the number c in Theorem 4.1 by tracing its proof". However, neither Knudsen et al. nor Bruns and Gubeladze do this. Part of the reason is that the bound obtained "by tracing the proof" would certainly be a tower of exponentials of length at least the initial maximal volume V.

This is because applying the result for a given initial maximal volume V yields a triangulation with maximal volume at most $V-1$, but we do not have easy control on the new number N' of simplices of volume $V-1$. Using Lemma 4.17 for this would give $N' = c^{Nd}\operatorname{vol}(P)$. Then, in the second step we then get that our bound for the number of cells of volume $V-2$ is $c^{c^{Nd}\operatorname{vol}(P)d}\operatorname{vol}(P)$, in the third step we get $c^{c^{c^{Nd}\operatorname{vol}(P)d}\operatorname{vol}(P)d}\operatorname{vol}(P)$, and so on.

Observe that if all simplices of maximal volume in $\mathcal{T}$ have non-degenerate box points whose carriers have disjoint stars then we only need to apply Corollary 4.18 once to get a triangulation with smaller maximal volume, leading to a bound of c instead of c^N in Theorem 4.19. If, moreover, the disjoint-stars property could be preserved in the iteration, we would get an effective KMW Theorem with a singly exponential factor of type $(d+1)!V$, rather than the double exponential of our Theorem 4.5.

Unfortunately, the hypothesis that stars are disjoint is not always satisfied and seems not easy to be preserved. However, it automatically holds when V is a prime.

LEMMA 4.21. *If Δ is a lattice simplex of volume V with respect to a certain lattice Λ, then:*

(1) *Δ has a unique minimal face F_0 of volume V.*
(2) *If V is prime, then every (non-zero) box point of F_0 is non-degenerate.*

PROOF. Let F_1 and F_2 be two faces of volume V of Δ. F_1 and F_2 cannot be disjoint, because the volume of a join is at least the product of the volumes. So, let $F_3 = F_1 \cap F_2$. Then either F_3 also has volume V or Δ has volume greater than V. More precisely, let Λ_i, $i = 1, 2, 3$, be the sublattice of Λ in the affine span of F_i. Then, the volume of Δ in the sublattice $\Lambda_1 + \Lambda_2$ times the volume of F_3 in Λ_3 equals the volume of F_1 in Λ_1 times that of F_2 in Λ_2. This proves part (1).

For part (2), observe that V being prime implies that F_0 is the intersection of all the faces of Δ that are not unimodular. In particular, all proper faces of F_0 are unimodular and every box point of F_0 is non-degenerate. □

COROLLARY 4.22. *If $\mathcal{T}$ is a lattice simplicial complex in which no simplex has volume larger than V, then the minimal faces of volume V in $\mathcal{T}$ have disjoint stars. Further, if V is prime, these minimal faces of volume V have non-degenerate box points.* □

COROLLARY 4.23. *Let $\mathcal{T}$ be a lattice triangulation with all its faces of volume at most V, for a prime V. Then, for every $c \in (d+1)!\mathbb{N}$, the dilation $c\mathcal{T}$ has a regular refinement in which every cell has volume less than V.*

PROOF. Let $F_1, \ldots, F_N$ be the list of minimal faces of volume V in $\mathcal{T}$. By Corollary 4.22, they have disjoint stars and possess non-degenerate box points $\boldsymbol{m}_1, \ldots, \boldsymbol{m}_N$. We then satisfy the conditions of Corollary 4.18, with the added feature that every simplex of volume V is in one of the stars of the F_i's. □

4.5. An effective version of the KMW-Theorem

Our proof of Theorem 4.19 (from which Theorem 4.1 follows) already departs from the previous proofs in that we use the "canonical triangulation" structure of the pieces after each refinement in order to control the number of iterations needed to decrease the maximum volume of simplices. This avoids us the use of "local lattices" [**25**] or "rational structures" [**77**], and it gives a cleaner proof of regularity.

Here we push this idea further and relate by an "**A**-structure" simplices that do not come from the canonical triangulation of the same dilated simplex. As usual, we let L_Δ denote the (affine) lattice generated by the vertices of a simplex Δ, and say that two full-dimensional ordered lattice simplices $\Delta = \operatorname{conv}\{\boldsymbol{a}_0, \ldots, \boldsymbol{a}_d\}$ and $\Delta' = \operatorname{conv}\{\boldsymbol{b}_0, \ldots, \boldsymbol{b}_d\}$ are **A**-*equivalent* if $L_\Delta = L_{\Delta'}$ and

$$\{\boldsymbol{a}_i - \boldsymbol{a}_{i-1} : i \in [d]\} = \{\boldsymbol{b}_i - \boldsymbol{b}_{i-1} : i \in [d]\}.$$

Lemma 4.24 offers ties between the canonical triangulation of $c\Delta$ and **A**-equivalence.

LEMMA 4.24. (1) *All the simplices in the canonical triangulation of $c\Delta$ are **A**-equivalent to Δ.*
(2) *If two simplices Δ and Δ' are **A**-equivalent then the **A**-dicing defined by Δ and by Δ' are the same, modulo a translation in Λ.* □

The second part of Lemma 4.24 allows us to understand a box point $\boldsymbol{m}$ for a simplex Δ as a box point for any **A**-equivalent simplex Δ'. Specifically, let $\boldsymbol{m}'$ be the translation of $\boldsymbol{m}$ obtained by sending the **A**-dicing of Δ to that of Δ'. This translation is unique, modulo the linear lattice $\vec{L}_\Delta$. The key property of **A**-equivalence that we need is the following:

LEMMA 4.25. *Let Δ and Δ' be two **A**-equivalent full-dimensional simplices in a triangulation $\mathcal{T}$, and $\boldsymbol{m}$ be a box point for Δ and $\boldsymbol{m}'$ the corresponding box point for Δ'. If F and F' be the carrier faces of $\boldsymbol{m}$ and $\boldsymbol{m}'$ in Δ and Δ' (that is, F and F' are the faces for which $\boldsymbol{m}$ and $\boldsymbol{m}'$ are non-degenerate box points), then either $F = F'$ or $\operatorname{conv}(F \cup F')$ is not a simplex in $\mathcal{T}$.*

PROOF. Although we defined box points as equivalent classes, let us now think of $\boldsymbol{m}$ and $\boldsymbol{m}'$ as representatives for their classes. That is, $\boldsymbol{m}$ and $\boldsymbol{m}'$ are two lattice points in the ambient lattice Λ.

Let F_0 be the minimal face of the **A**-dicing of Δ containing $\boldsymbol{m}$. F_0 might not be equal to F, since F_0 may not be a face of Δ, but without loss of generality (by changing the representative for $\boldsymbol{m}$ if needed) we can assume that $\operatorname{aff}(F_0) \subset \operatorname{aff}(F)$. (This is because the flat spanned by F contains representatives for $\boldsymbol{m}$, and the **A**-dicing induced by Δ is a hyperplane arrangement refining the flat spanned by every face of Δ). For the same reason, we assume that the minimal face F_0', of the **A**-dicing of Δ' containing $\boldsymbol{m}'$, satisfies $\operatorname{aff}(F_0') \subset \operatorname{aff}(F')$.

Suppose now that $\operatorname{conv}(F \cup F')$ is a simplex in $\mathcal{T}$. Since $\operatorname{aff}(F_0)$ and $\operatorname{aff}(F_0')$ are parallel flats contained in respective faces of $\operatorname{conv}(F \cup F')$, we must have $\operatorname{aff}(F_0) = \operatorname{aff}(F_0')$, so $F \cap F'$ is not empty. In particular, the **A**-dicings of Δ and Δ' are not only translations of one another, but actually equal. Hence, $\boldsymbol{m} = \boldsymbol{m}'$, modulo the (linear) lattice $\vec{L}_\Delta = \vec{L}'_\Delta$.

By Lemma 4.21, there is a unique face of $\operatorname{conv}(F \cup F')$ for which $\boldsymbol{m} = \boldsymbol{m}'$ is a non-degenerate box point. That face must be $F = F'$. □

Hence, if we choose an **A**-class of simplices of a triangulation $\mathcal{T}$, Corollary 4.18 can be applied simultaneously to faces whose stars contain all the simplices in that class.

COROLLARY 4.26. *Let $\mathcal{T}$ be a triangulation with a total order in its vertices. If $\Delta_1, \ldots, \Delta_N$ are representatives for all the **A**-equivalence classes of full-dimensional simplices in $\mathcal{T}$, and $V_1, \ldots, V_N$ are their respective volumes, then the total number of **A**-equivalence classes of simplices used to obtain a unimodular triangulation by iterative applications of Theorem 4.19 is bounded above by:*

$$\sum_{i=1}^{N} V_i! \left((d+1)!c^d\right)^{V_i - 1}$$

PROOF. The statement follows from the claim that the number of **A**-equivalence classes that arise in the refinements of all the dilations of simplices of a particular **A**-equivalence class of volume V is bounded above by

$$V! \left((d+1)!c^d\right)^{V-1}.$$

We prove this claim via induction on V, starting with the case $V = 1$. In this case the number of classes is one. Now consider the case where $V > 1$, and take a particular Δ in a certain **A**-equivalence class. Once its box point $\boldsymbol{m}$ is chosen, $c\Delta$

gets refined, by Lemma 4.17, into at most $(d+1)c^d$ simplices of volume at most $V-1$. By inductive hypothesis each of them will produce at most

$$(V-1)!\,\left((d+1)!c^d\right)^{V-2}$$

A-classes when further and further refined. So, the total number of new classes produced by refining Δ is bounded by

$$\begin{aligned} &(d+1)c^d(V-1)!\,\left((d+1)!c^d\right)^{V-2} \\ = &\frac{V!\,\left((d+1)!c^d\right)^{V-1}}{Vd!}. \end{aligned}$$

Now, it is not true that all choices of Δ in the same **A**-class will produce the same **A**-classes when refined. On the one hand, we must take into account that there are $d!$ translation classes within each **A**-class. On the other hand, simplices Δ and Δ' that are translations of one another may get different refinements, because the refinement depends on the choice of the box point $\boldsymbol{m}$. But these two (translation class and box point) are the only things that can make the dilations of two **A**-equivalent simplices have canonical refinements that are not **A**-equivalent to one another.

Since there are $V-1$ choices for $\boldsymbol{m}$, the total number of classes produced from an individual class of volume V (including the initial class itself) is bounded above by

$$\begin{aligned} &1+d!(V-1)\frac{V!\,\left((d+1)!c^d\right)^{V-1}}{Vd!} \\ < \; &V!\,\left((d+1)!c^d\right)^{V-1}. \end{aligned}$$

□

This implies the following, which in turn proves Theorem 4.5:

THEOREM 4.27. *Let $\mathcal{T}$ be a triangulation of a lattice polytope P. In the notation of Corollary 4.26, the dilation $c^M\mathcal{T}$ has a regular unimodular refinement for every $c \in (d+1)!\mathbb{N}$ and for every $M \geq \sum_{i=1}^{N} V_i!\,\left((d+1)!c^d\right)^{V_i-1}$.*

PROOF. Apply Corollary 4.18 to each **A**-class of simplices in $\mathcal{T}$ (and the new ones created in the process) starting with those of higher volume. At each iteration, consider box points and faces whose stars completely cover one **A**-class of maximal volume, which can be done by Lemma 4.25. Since that class will not appear in future steps (because only classes of strictly smaller volume can appear), the total number of iterations is bounded above by the total number of **A**-classes that can arise in the process, which is bounded by M according to Corollary 4.26. □

REMARK 4.28. This proof does not claim that all the simplices of a particular class that appear anywhere in the process are refined in the same step. Some may get refined "before their turn" if they are in the stars of the faces F_i used to refine other classes. But this does not invalidate the proof. We devote one iteration to addressing whatever is left from each particular class, so the number of iterations is at most the number of classes.

Bibliography

[1] 4ti2 team, `4ti2` — *a software package for algebraic, geometric and combinatorial problems on linear spaces*, Available at `http://www.4ti2.de`.

[2] Karen Aardal, Robert Weismantel, and Laurence A. Wolsey, *Non-standard approaches to integer programming*, Discrete Appl. Math. **123** (2002), no. 1-3, 5–74, DOI 10.1016/S0166-218X(01)00337-7. Workshop on Discrete Optimization, DO'99 (Piscataway, NJ). MR1922330

[3] Per Alexandersson, *Gelfand-Tsetlin polytopes and the integer decomposition property*, European J. Combin. **54** (2016), 1–20, DOI 10.1016/j.ejc.2015.11.006. MR3459049

[4] Satoshi Aoki, Takayuki Hibi, Hidefumi Ohsugi, and Akimichi Takemura, *Gröbner bases of nested configurations*, J. Algebra **320** (2008), no. 6, 2583–2593, DOI 10.1016/j.jalgebra.2008.05.023. MR2441772

[5] V. I. Arnol′d, *Statistics of integral convex polygons* (Russian), Funktsional. Anal. i Prilozhen. **14** (1980), no. 2, 1–3. MR575199

[6] Christos A. Athanasiadis, *Ehrhart polynomials, simplicial polytopes, magic squares and a conjecture of Stanley*, J. Reine Angew. Math. **583** (2005), 163–174, DOI 10.1515/crll.2005.2005.583.163. MR2146855

[7] I. Bárány and A. M. Vershik, *On the number of convex lattice polytopes*, Geom. Funct. Anal. **2** (1992), no. 4, 381–393, DOI 10.1007/BF01896660. MR1191566

[8] I. Bárány and L. Yuan, *Volumes of convex lattice polytopes and a question of V. I. Arnold*, Acta Math. Hungar. **144** (2014), no. 1, 119–131, DOI 10.1007/s10474-014-0418-0. MR3267174

[9] Alexander Barvinok, *A course in convexity*, Graduate Studies in Mathematics, vol. 54, American Mathematical Society, Providence, RI, 2002, DOI 10.1090/gsm/054. MR1940576

[10] Victor V. Batyrev, *Non-Archimedean integrals and stringy Euler numbers of log-terminal pairs*, J. Eur. Math. Soc. (JEMS) **1** (1999), no. 1, 5–33, DOI 10.1007/PL00011158. MR1677693

[11] Victor V. Batyrev and Dimitrios I. Dais, *Strong McKay correspondence, string-theoretic Hodge numbers and mirror symmetry*, Topology **35** (1996), no. 4, 901–929, DOI 10.1016/0040-9383(95)00051-8. MR1404917

[12] Victor V. Batyrev and Johannes Hofscheier, *A generalization of a theorem of gk white*, arXiv preprint arXiv:1004.3411 (2010), 1–12.

[13] Matthias Beck, Benjamin Braun, Matthias Köppe, Carla D. Savage, and Zafeirakis Zafeirakopoulos, *Generating functions and triangulations for lecture hall cones*, SIAM J. Discrete Math. **30** (2016), no. 3, 1470–1479, DOI 10.1137/15M1036907. MR3531728

[14] Matthias Beck, Beifang Chen, Lenny Fukshansky, Christian Haase, Allen Knutson, Bruce Reznick, Sinai Robins, and Achill Schürmann, *Problems from the Cottonwood Room*, Integer points in polyhedra—geometry, number theory, algebra, optimization, Contemp. Math., vol. 374, Amer. Math. Soc., Providence, RI, 2005, pp. 179–191, DOI 10.1090/conm/374/06905. MR2134767

[15] Matthias Beck and Sinai Robins, *Computing the continuous discretely*, 2nd ed., Undergraduate Texts in Mathematics, Springer, New York, 2015. Integer-point enumeration in polyhedra; With illustrations by David Austin, DOI 10.1007/978-1-4939-2969-6. MR3410115

[16] Matthias Beck and Thomas Zaslavsky, *Inside-out polytopes*, Adv. Math. **205** (2006), no. 1, 134–162, DOI 10.1016/j.aim.2005.07.006. MR2254310

[17] A. D. Berenstein and A. V. Zelevinsky, *Tensor product multiplicities and convex polytopes in partition space*, J. Geom. Phys. **5** (1988), no. 3, 453–472, DOI 10.1016/0393-0440(88)90033-2. MR1048510

[18] Arkady Berenstein and Andrei Zelevinsky, *Tensor product multiplicities, canonical bases and totally positive varieties*, Invent. Math. **143** (2001), no. 1, 77–128, DOI 10.1007/s002220000102. MR1802793

[19] U. Betke and P. McMullen, *Lattice points in lattice polytopes*, Monatsh. Math. **99** (1985), no. 4, 253–265, DOI 10.1007/BF01312545. MR799674

[20] Mireille Bousquet-Mélou and Kimmo Eriksson, *Lecture hall partitions*, Ramanujan J. **1** (1997), no. 1, 101–111, DOI 10.1023/A:1009771306380. MR1607531

[21] Srinivasa Ramanujan, *Highly composite numbers*, Ramanujan J. **1** (1997), no. 2, 119–153, DOI 10.1023/A:1009764017495. Annotated and with a foreword by Jean-Louis Nicolas and Guy Robin. MR1606180

[22] Petter Brändén and Madeleine Leander, *Lecture hall p-partitions*, J. Combin. **11** (2020), pp. 391–412, DOI: 10.4310/JOC.2020.v11.n2.a9; arxiv:1609.02790, pp. 1–15, 2016.

[23] Felix Breuer and Aaron Dall, *Bounds on the coefficients of tension and flow polynomials*, J. Algebraic Combin. **33** (2011), no. 3, 465–482, DOI 10.1007/s10801-010-0254-4. MR2772543

[24] Felix Breuer and Raman Sanyal, *Ehrhart theory, modular flow reciprocity, and the Tutte polynomial*, Math. Z. **270** (2012), no. 1-2, 1–18, DOI 10.1007/s00209-010-0782-6. MR2875820

[25] Winfried Bruns and Joseph Gubeladze, *Normality and covering properties of affine semigroups*, J. Reine Angew. Math. **510** (1999), 161–178, DOI 10.1515/crll.1999.044. MR1696094

[26] Winfried Bruns and Joseph Gubeladze, *Unimodular covers of multiples of polytopes*, Doc. Math. **7** (2002), 463–480. MR2014490

[27] Winfried Bruns and Joseph Gubeladze, *Polytopes, rings, and K-theory*, Springer Monographs in Mathematics, Springer, Dordrecht, 2009, DOI 10.1007/b105283. MR2508056

[28] Winfried Bruns, Joseph Gubeladze, and Ngô Viêt Trung, *Normal polytopes, triangulations, and Koszul algebras*, J. Reine Angew. Math. **485** (1997), 123–160. MR1442191

[29] Winfried Bruns, Joseph Gubeladze, and Ngô Viêt Trung, *Problems and algorithms for affine semigroups*, Semigroup Forum **64** (2002), no. 2, 180–212, DOI 10.1007/s002330010099. MR1876854

[30] Winfried Bruns, Jürgen Herzog, and Udo Vetter, *Syzygies and walks*, Commutative algebra (Trieste, 1992), World Sci. Publ., River Edge, NJ, 1994, pp. 36–57. MR1421076

[31] Winfried Bruns and Bogdan Ichim, *Normaliz: algorithms for affine monoids and rational cones*, J. Algebra **324** (2010), no. 5, 1098–1113, DOI 10.1016/j.jalgebra.2010.01.031. MR2659215

[32] Winfried Bruns and Tim Römer, *h-vectors of Gorenstein polytopes*, J. Combin. Theory Ser. A **114** (2007), no. 1, 65–76, DOI 10.1016/j.jcta.2006.03.003. MR2275581

[33] Winfried Bruns and Christof Söger, *The computation of generalized Ehrhart series in Normaliz.* part 2, J. Symbolic Comput. **68** (2015), no. part 2, 75–86, DOI 10.1016/j.jsc.2014.09.004. MR3283855

[34] Anders Skovsted Buch, *The saturation conjecture (after A. Knutson and T. Tao)*, Enseign. Math. (2) **46** (2000), no. 1-2, 43–60. With an appendix by William Fulton. MR1769536

[35] Weronika Buczyńska and Jarosław A. Wiśniewski, *On geometry of binary symmetric models of phylogenetic trees*, J. Eur. Math. Soc. (JEMS) **9** (2007), no. 3, 609–635, DOI 10.4171/JEMS/90. MR2314109

[36] Thomas Chappell, Tobias Friedl, and Raman Sanyal, *Two double poset polytopes*, SIAM J. Discrete Math. **31** (2017), no. 4, 2378–2413, DOI 10.1137/16M1091800. MR3715376

[37] Soojin Cho, *Polytopes of roots of type A_n*, Bull. Austral. Math. Soc. **59** (1999), no. 3, 391–402, DOI 10.1017/S0004972700033062. MR1697418

[38] V. Chvátal, *On certain polytopes associated with graphs*, J. Combinatorial Theory Ser. B **18** (1975), 138–154, DOI 10.1016/0095-8956(75)90041-6. MR0371732

[39] CoCoATeam, CoCoA*: a system for doing Computations in Commutative Algebra*, Available at http://cocoa.dima.unige.it.

[40] David A. Cox, Christian Haase, Takayuki Hibi, and Akihiro Higashitani, *Integer decomposition property of dilated polytopes*, Electron. J. Combin. **21** (2014), no. 4, Paper 4.28, 17. MR3292265

[41] Jesús A. De Loera, *The many aspects of counting lattice points in polytopes*, Math. Semesterber. **52** (2005), no. 2, 175–195, DOI 10.1007/s00591-005-0094-9. MR2159956

[42] Jesús A. De Loera, Raymond Hemmecke, and Matthias Köppe, *Algebraic and geometric ideas in the theory of discrete optimization*, MOS-SIAM Series on Optimization, vol. 14,

Society for Industrial and Applied Mathematics (SIAM), Philadelphia, PA; Mathematical Optimization Society, Philadelphia, PA, 2013. MR3024570

[43] Jesús A. De Loera and Tyrrell B. McAllister, *On the computation of Clebsch-Gordan coefficients and the dilation effect*, Experiment. Math. **15** (2006), no. 1, 7–19. MR2229381

[44] Jesús A. De Loera, Jörg Rambau, and Francisco Santos, *Triangulations*, Algorithms and Computation in Mathematics, vol. 25, Springer-Verlag, Berlin, 2010. Structures for algorithms and applications, DOI 10.1007/978-3-642-12971-1. MR2743368

[45] Wolfram Decker, Gert-Martin Greuel, Günter Pfister, and Hans Schönemann, SINGULAR *3.1.6 — A computer algebra system for polynomial computations*, `www.singular.uni-kl.de`, 2012.

[46] Persi Diaconis and Nicholas Eriksson, *Markov bases for noncommutative Fourier analysis of ranked data*, J. Symbolic Comput. **41** (2006), no. 2, 182–195, DOI 10.1016/j.jsc.2005.04.009. MR2197154
Math. **227** (1967), 25–49. MR36#105

[47] E. Ehrhart, *Sur un problème de géométrie diophantienne linéaire. II. Systèmes diophantiens linéaires* (French), J. Reine Angew. Math. **227** (1967), 25–49, DOI 10.1515/crll.1967.227.25. MR0217010

[48] David Eisenbud, Alyson Reeves, and Burt Totaro, *Initial ideals, Veronese subrings, and rates of algebras*, Adv. Math. **109** (1994), no. 2, 168–187, DOI 10.1006/aima.1994.1085. MR1304751

[49] R. M. Erdahl and S. S. Ryshkov, *On lattice dicing*, European J. Combin. **15** (1994), no. 5, 459–481, DOI 10.1006/eujc.1994.1049. MR1292957

[50] Günter Ewald, *Combinatorial convexity and algebraic geometry*, Graduate Texts in Mathematics, vol. 168, Springer-Verlag, New York, 1996, DOI 10.1007/978-1-4612-4044-0. MR1418400

[51] R. T. Firla and G. M. Ziegler, *Hilbert bases, unimodular triangulations, and binary covers of rational polyhedral cones*, Discrete Comput. Geom. **21** (1999), no. 2, 205–216, DOI 10.1007/PL00009416. MR1668094

[52] Dominique Foata, *Distributions eulériennes et mahoniennes sur le groupe des permutations* (French), Higher combinatorics (Proc. NATO Advanced Study Inst., Berlin, 1976), NATO Adv. Study Inst. Ser., Ser. C: Math. Phys. Sci., vol. 31, Reidel, Dordrecht-Boston, Mass., 1977, pp. 27–49. With a comment by Richard P. Stanley. MR519777

[53] William Fulton, *Introduction to toric varieties*, Annals of Mathematics Studies, vol. 131, Princeton University Press, Princeton, NJ, 1993. The William H. Roever Lectures in Geometry, DOI 10.1515/9781400882526. MR1234037

[54] Ewgenij Gawrilow and Michael Joswig, *polymake: a framework for analyzing convex polytopes*, Polytopes—combinatorics and computation (Oberwolfach, 1997), DMV Sem., vol. 29, Birkhäuser, Basel, 2000, pp. 43–73. MR1785292

[55] Israel M. Gelfand, Mark I. Graev, and Alexander Postnikov, *Combinatorics of hypergeometric functions associated with positive roots*, The Arnold-Gelfand mathematical seminars, Birkhäuser Boston, Boston, MA, 1997, pp. 205–221, DOI 10.1007/978-1-4612-4122-5_10. MR1429893

[56] I. M. Gel′fand, M. M. Kapranov, and A. V. Zelevinsky, *Discriminants, resultants, and multidimensional determinants*, Mathematics: Theory & Applications, Birkhäuser Boston, Inc., Boston, MA, 1994, DOI 10.1007/978-0-8176-4771-1. MR1264417

[57] Daniel R. Grayson and Michael E. Stillman, *Macaulay2, a software system for research in algebraic geometry*, Available at http://www.math.uiuc.edu/Macaulay2/.

[58] Joseph Gubeladze, *Convex normality of rational polytopes with long edges*, Adv. Math. **230** (2012), no. 1, 372–389, DOI 10.1016/j.aim.2011.12.003. MR2900547

[59] Christian Haase, Benjamin Lorenz, and Andreas Paffenholz, *Generating smooth lattice polytopes*, Mathematical software—ICMS 2010, Lecture Notes in Comput. Sci., vol. 6327, Springer, Berlin, 2010, pp. 315–328. MR3663208

[60] Christian Haase and Andreas Paffenholz, *On Fanos and chimneys*, Oberwolfach Rep. **4** (2007), no. 3, 2303–2306.

[61] Christian Haase and Andreas Paffenholz, *Quadratic Gröbner bases for smooth* 3×3 *transportation polytopes*, J. Algebraic Combin. **30** (2009), no. 4, 477–489, DOI 10.1007/s10801-009-0173-4. MR2563137

[62] Martin Henk and Robert Weismantel, *The height of minimal Hilbert bases*, Results Math. **32** (1997), no. 3-4, 298–303, DOI 10.1007/BF03322141. MR1487601

[63] Patricia Hersh and Ed Swartz, *Coloring complexes and arrangements*, J. Algebraic Combin. **27** (2008), no. 2, 205–214, DOI 10.1007/s10801-007-0086-z. MR2375492

[64] Jürgen Herzog, Takayuki Hibi, and Hidefumi Ohsugi, *Unmixed bipartite graphs and sublattices of the Boolean lattices*, J. Algebraic Combin. **30** (2009), no. 4, 415–420, DOI 10.1007/s10801-009-0168-1. MR2563133

[65] Takayuki Hibi, Akihiro Higashitani, and Koutarou Yoshida, *Existence of regular unimodular triangulations of dilated empty simplices*, European J. Combin. **70** (2018), 374–383, DOI 10.1016/j.ejc.2018.01.012. MR3779625

[66] James E. Humphreys, *Reflection groups and Coxeter groups*, Cambridge Studies in Advanced Mathematics, vol. 29, Cambridge University Press, Cambridge, 1990, DOI 10.1017/CBO9780511623646. MR1066460

[67] Yukari Ito and Miles Reid, *The McKay correspondence for finite subgroups of* $\mathrm{SL}(3, \mathbf{C})$, Higher-dimensional complex varieties (Trento, 1994), de Gruyter, Berlin, 1996, pp. 221-240. MR1463181

[68] Michael Joswig, *Essentials of tropical combinatorics*, Springer, Berlin, 2014.

[69] Michael Joswig, Benjamin Müller, and Andreas Paffenholz, `polymake` *and lattice polytopes*, 21st International Conference on Formal Power Series and Algebraic Combinatorics (FPSAC 2009), Discrete Math. Theor. Comput. Sci. Proc., AK, Assoc. Discrete Math. Theor. Comput. Sci., Nancy, 2009, pp. 491–502. MR2721537

[70] Volker Kaibel and Martin Wolff, *Simple 0/1-polytopes*, European J. Combin. **21** (2000), no. 1, 139–144, DOI 10.1006/eujc.1999.0328. Combinatorics of polytopes. MR1737334

[71] J.-M. Kantor and K. S. Sarkaria, *On primitive subdivisions of an elementary tetrahedron*, Pacific J. Math. **211** (2003), no. 1, 123–155, DOI 10.2140/pjm.2003.211.123. MR2016594

[72] M. M. Kapranov, B. Sturmfels, and A. V. Zelevinsky, *Chow polytopes and general resultants*, Duke Math. J. **67** (1992), no. 1, 189–218, DOI 10.1215/S0012-7094-92-06707-X. MR1174606

[73] Lars Kastner, Benjamin Lorenz, Andreas Paffenholz, and Anna-Lena Winz, *polymake_toric (extension to polymake)*, mar 2013, software.

[74] Lars Kastner, Benjamin Lorenz, and Anna-Lena Winz, *polymake_algebra (extension to polymake)*, mar 2013.

[75] G. Kempf, Finn Faye Knudsen, D. Mumford, and B. Saint-Donat, *Toroidal embeddings. I*, Lecture Notes in Mathematics, Vol. 339, Springer-Verlag, Berlin-New York, 1973. MR0335518

[76] Tomonori Kitamura, *Gröbner bases associated with positive roots and Catalan numbers*, Osaka J. Math. **42** (2005), no. 2, 421–433. MR2147728

[77] G. Kempf, Finn Faye Knudsen, D. Mumford, and B. Saint-Donat, *Toroidal embeddings. I*, Lecture Notes in Mathematics, Vol. 339, Springer-Verlag, Berlin-New York, 1973. MR0335518

[78] Allen Knutson and Terence Tao, *The honeycomb model of* $\mathrm{GL}_n(\mathbf{C})$ *tensor products. I. Proof of the saturation conjecture*, J. Amer. Math. Soc. **12** (1999), no. 4, 1055–1090, DOI 10.1090/S0894-0347-99-00299-4. MR1671451

[79] Michał Lasoń and Mateusz Michałek, *On the toric ideal of a matroid*, Adv. Math. **259** (2014), 1–12, DOI 10.1016/j.aim.2014.03.004. MR3197649

[80] Michał Lasoń and Mateusz Michałek, *Non-normal very ample polytopes—constructions and examples*, Exp. Math. **26** (2017), no. 2, 130–137, DOI 10.1080/10586458.2015.1128370. MR3623864

[81] Carl W. Lee, *Subdivisions and triangulations of polytopes*, Handbook of discrete and computational geometry, CRC Press Ser. Discrete Math. Appl., CRC, Boca Raton, FL, 1997, pp. 271–290. MR1730170

[82] Matthias Lenz, *Torische Ideale von Flusspolytopen*, Master's thesis, FU Berlin, November 2007, `arxiv:`0709.3570.

[83] Fu Liu and Richard P. Stanley, *The lecture hall parallelepiped*, Ann. Comb. **18** (2014), no. 3, 473–488, DOI 10.1007/s00026-014-0235-8. MR3245894

[84] Anders Lundman, *A classification of smooth convex 3-polytopes with at most 16 lattice points*, J. Algebraic Combin. **37** (2013), no. 1, 139–165, DOI 10.1007/s10801-012-0363-3. MR3016305

[85] D. Maclagan and R. R. Thomas, *Combinatorics of the toric Hilbert scheme*, Discrete Comput. Geom. **27** (2002), no. 2, 249–272, DOI 10.1007/s00454-001-0062-7. MR1880941

[86] Karola Mészáros, *Root polytopes, triangulations, and the subdivision algebra. I*, Trans. Amer. Math. Soc. **363** (2011), no. 8, 4359–4382, DOI 10.1090/S0002-9947-2011-05265-7. MR2792991

[87] Karola Mészáros, *Root polytopes, triangulations, and the subdivision algebra, II*, Trans. Amer. Math. Soc. **363** (2011), no. 11, 6111–6141, DOI 10.1090/S0002-9947-2011-05371-7. MR2817421

[88] *Mini-workshop: Ehrhart Quasipolynomials: Algebra, Combinatorics, and Geometry*, Oberwolfach Rep. **1** (2004), no. 3, 2071–2101, DOI 10.4171/OWR/2004/39. Abstracts from the mini-workshop held August 15–21, 2004; Organized by Jesús A. De Loera and Christian Haase; Oberwolfach Reports. Vol. 1, no. 3. MR2144157

[89] *Mini-Workshop: Projective Normality of Smooth Toric Varieties*, Oberwolfach Rep. **4** (2007), no. 3, 2283–2319. Abstracts from the mini-workshop held August 12–18, 2007; Organized by Christian Haase, Takayuki Hibi and Diane Maclagan; Oberwolfach Reports. Vol. 4, no. 3. MR2432116

[90] David R. Morrison and Glenn Stevens, *Terminal quotient singularities in dimensions three and four*, Proc. Amer. Math. Soc. **90** (1984), no. 1, 15–20, DOI 10.2307/2044659. MR722406

[91] Mircea Mustaţă and Sam Payne, *Ehrhart polynomials and stringy Betti numbers*, Math. Ann. **333** (2005), no. 4, 787–795, DOI 10.1007/s00208-005-0691-x. MR2195143

[92] Mikkel Øbro, *An algorithm for the classification of smooth fano polytopes*, preprint, arXiv:0704.0049, 2007.

[93] Tadao Oda, *Convex bodies and algebraic geometry*, Ergebnisse der Mathematik und ihrer Grenzgebiete (3) [Results in Mathematics and Related Areas (3)], vol. 15, Springer-Verlag, Berlin, 1988. An introduction to the theory of toric varieties; Translated from the Japanese. MR922894

[94] Hidefumi Ohsugi, *Toric ideals and an infinite family of normal* $(0, 1)$*-polytopes without unimodular regular triangulations*, Discrete Comput. Geom. **27** (2002), no. 4, 551–565, DOI 10.1007/s00454-001-0087-y. MR1902677

[95] Hidefumi Ohsugi and Takayuki Hibi, *Normal polytopes arising from finite graphs*, J. Algebra **207** (1998), no. 2, 409–426, DOI 10.1006/jabr.1998.7476. MR1644250

[96] Hidefumi Ohsugi and Takayuki Hibi, *Koszul bipartite graphs*, Adv. in Appl. Math. **22** (1999), no. 1, 25–28, DOI 10.1006/aama.1998.0615. MR1657721

[97] H. Ohsugi and T. Hibi, *A normal* $(0, 1)$*-polytope none of whose regular triangulations is unimodular*, Discrete Comput. Geom. **21** (1999), no. 2, 201–204, DOI 10.1007/PL00009415. MR1668090

[98] Hidefumi Ohsugi and Takayuki Hibi, *Toric ideals generated by quadratic binomials*, J. Algebra **218** (1999), no. 2, 509–527, DOI 10.1006/jabr.1999.7918. MR1705794

[99] Hidefumi Ohsugi and Takayuki Hibi, *Convex polytopes all of whose reverse lexicographic initial ideals are squarefree*, Proc. Amer. Math. Soc. **129** (2001), no. 9, 2541–2546, DOI 10.1090/S0002-9939-01-05853-1. MR1838375

[100] Hidefumi Ohsugi and Takayuki Hibi, *Unimodular triangulations and coverings of configurations arising from root systems*, J. Algebraic Combin. **14** (2001), no. 3, 199–219, DOI 10.1023/A:1012772002661. MR1869410

[101] Hidefumi Ohsugi and Takayuki Hibi, *Quadratic initial ideals of root systems*, Proc. Amer. Math. Soc. **130** (2002), no. 7, 1913–1922, DOI 10.1090/S0002-9939-01-06411-5. MR1896022

[102] ______, *Simple polytopes arising from finite graphs*, Proceedings of the 2008 International Conference on Information Theory and Statistical Learning, ITSL 2008, CSREA Press, pp. 73–79, 2008; `arxiv:0804.4287`, April 2008.

[103] Hidefumi Ohsugi and Takayuki Hibi, *Toric rings and ideals of nested configurations*, J. Commut. Algebra **2** (2010), no. 2, 187–208, DOI 10.1216/JCA-2010-2-2-187. MR2647475

[104] Andreas Paffenholz, *Projections of subdivisions, an extension for* `polymake`, available at github:apaffenholz/polymake_projectionWithSubdivision, March 2012.

[105] Igor Pak, *Hook length formula and geometric combinatorics*, Sém. Lothar. Combin. **46** (2001/02), Art. B46f, 13. MR1877632

[106] Sam Payne, *Frobenius splittings of toric varieties*, Algebra Number Theory **3** (2009), no. 1, 107–119, DOI 10.2140/ant.2009.3.107. MR2491910

[107] Irena Peeva and Mike Stillman, *Toric Hilbert schemes*, Duke Math. J. **111** (2002), no. 3, 419–449, DOI 10.1215/S0012-7094-02-11132-6. MR1885827

[108] Georg Alexander Pick, *Geometrisches zur Zahlenlehre*, Sitzenber. Lotos (Prague) **19** (1899), 311–319.

[109] polymake team, *polymake version 2.12*, available at http://www.polymake.org, March 2012.

[110] Alexander Postnikov, *Permutohedra, associahedra, and beyond*, Int. Math. Res. Not. IMRN **6** (2009), 1026–1106, DOI 10.1093/imrn/rnn153. MR2487491

[111] Jörg Rambau, *TOPCOM: triangulations of point configurations and oriented matroids*, Mathematical software (Beijing, 2002), World Sci. Publ., River Edge, NJ, 2002, pp. 330–340. MR1932619
Soc. **7** (1957), 378–395.

[112] J. E. Reeve, *On the volume of lattice polyhedra*, Proc. London Math. Soc. (3) **7** (1957), 378–395, DOI 10.1112/plms/s3-7.1.378. MR0095452

[113] Miles Reid, *McKay correspondence*, Proc. of Algebraic Geometry Symposium (Kinosaki, Nov 1996), 1997, arxiv:math.AG/9702016, pp. 14–41.

[114] Francisco Santos, *A point set whose space of triangulations is disconnected*, J. Amer. Math. Soc. **13** (2000), no. 3, 611–637, DOI 10.1090/S0894-0347-00-00330-1. MR1758756

[115] Francisco Santos, *Non-connected toric Hilbert schemes*, Math. Ann. **332** (2005), no. 3, 645–665, DOI 10.1007/s00208-005-0643-5. MR2181765

[116] F. Santos and G. M. Ziegler, *Unimodular triangulations of dilated 3-polytopes*, Trans. Moscow Math. Soc., posted on 2013, 293–311, DOI 10.1090/s0077-1554-2014-00220-x. MR3235802

[117] Carla D. Savage, *The mathematics of lecture hall partitions*, J. Combin. Theory Ser. A **144** (2016), 443–475, DOI 10.1016/j.jcta.2016.06.006. MR3534075

[118] Carla D. Savage and Michael J. Schuster, *Ehrhart series of lecture hall polytopes and Eulerian polynomials for inversion sequences*, J. Combin. Theory Ser. A **119** (2012), no. 4, 850–870, DOI 10.1016/j.jcta.2011.12.005. MR2881231

[119] Herbert E. Scarf, *Integral polyhedra in three space*, Math. Oper. Res. **10** (1985), no. 3, 403–438, DOI 10.1287/moor.10.3.403. MR798388

[120] Alexander Schrijver, *Theory of linear and integer programming*, Wiley-Interscience Series in Discrete Mathematics, John Wiley & Sons, Ltd., Chichester, 1986. A Wiley-Interscience Publication. MR874114

[121] Alexander Schrijver, *Combinatorial optimization. Polyhedra and efficiency. Vol. A*, Algorithms and Combinatorics, vol. 24, Springer-Verlag, Berlin, 2003. Paths, flows, matchings. MR1956924

[122] András Sebő, *Hilbert bases, Carathéodory's theorem and combinatorial optimization*, Integer Programming and Combinatorial Optimization (Ravindran Kannan and William R. Pulleyblank, eds.), Math. Prog. Soc., Univ. Waterloo Press, 1990, pp. 431–456.

[123] Aron Simis, Wolmer V. Vasconcelos, and Rafael H. Villarreal, *On the ideal theory of graphs*, J. Algebra **167** (1994), no. 2, 389–416, DOI 10.1006/jabr.1994.1192. MR1283294

[124] Aron Simis, Wolmer V. Vasconcelos, and Rafael H. Villarreal, *The integral closure of subrings associated to graphs*, J. Algebra **199** (1998), no. 1, 281–289, DOI 10.1006/jabr.1997.7171. MR1489364

[125] Richard P. Stanley, *Eulerian partitions of a unit hypercube*, 1977, Appendix to [**52**].

[126] Richard P. Stanley, *Decompositions of rational convex polytopes*, Ann. Discrete Math. **6** (1980), 333–342. Combinatorial mathematics, optimal designs and their applications (Proc. Sympos. Combin. Math. and Optimal Design, Colorado State Univ., Fort Collins, Colo., 1978). MR593545

[127] Richard P. Stanley, *Two poset polytopes*, Discrete Comput. Geom. **1** (1986), no. 1, 9–23, DOI 10.1007/BF02187680. MR824105

[128] Einar Steingrímsson, *The coloring ideal and coloring complex of a graph*, J. Algebraic Combin. **14** (2001), no. 1, 73–84, DOI 10.1023/A:1011222121664. MR1856230

[129] Michael Stillman, Bernd Sturmfels, and Rekha Thomas, *Algorithms for the toric Hilbert scheme*, Computations in algebraic geometry with Macaulay 2, Algorithms Comput. Math., vol. 8, Springer, Berlin, 2002, pp. 179–214, DOI 10.1007/978-3-662-04851-1_8. MR1949552

[130] Bernd Sturmfels, *Gröbner bases of toric varieties*, Tohoku Math. J. (2) **43** (1991), no. 2, 249–261, DOI 10.2748/tmj/1178227496. MR1104431

[131] ———, *The geometry of A-graded algebras*, arxiv:math.AG/9410032, 1994.

[132] Bernd Sturmfels, *Gröbner bases and convex polytopes*, University Lecture Series, vol. 8, American Mathematical Society, Providence, RI, 1996. MR1363949

[133] Bernd Sturmfels and Seth Sullivant, *Toric geometry of cuts and splits*, Michigan Math. J. **57** (2008), 689–709, DOI 10.1307/mmj/1220879432. Special volume in honor of Melvin Hochster. MR2492476

[134] Seth Sullivant, *Compressed polytopes and statistical disclosure limitation*, Tohoku Math. J. (2) **58** (2006), no. 3, 433–445. MR2273279

[135] Seth Sullivant, *Toric fiber products*, J. Algebra **316** (2007), no. 2, 560–577, DOI 10.1016/j.jalgebra.2006.10.004. MR2356844

[136] Michael von Thaden, *Unimodular covers and triangulations of lattice polytopes*, Ph.D. thesis, Universität Osnabrück, 2007.

[137] G. K. White, *Lattice tetrahedra*, Canad. J. Math. **16** (1964), 389–396, DOI 10.4153/CJM-1964-040-2. MR0161837

[138] Neil L. White, *A unique exchange property for bases*, Linear Algebra Appl. **31** (1980), 81–91, DOI 10.1016/0024-3795(80)90209-8. MR570381

[139] Takashi Yamaguchi, Mitsunori Ogawa, and Akimichi Takemura, *Markov degree of the Birkhoff model*, J. Algebraic Combin. **40** (2014), no. 1, 293–311, DOI 10.1007/s10801-013-0488-z. MR3226827

Editorial Information

To be published in the *Memoirs*, a paper must be correct, new, nontrivial, and significant. Further, it must be well written and of interest to a substantial number of mathematicians. Piecemeal results, such as an inconclusive step toward an unproved major theorem or a minor variation on a known result, are in general not acceptable for publication.

Papers appearing in *Memoirs* are generally at least 80 and not more than 200 published pages in length. Papers less than 80 or more than 200 published pages require the approval of the Managing Editor of the Transactions/Memoirs Editorial Board. Published pages are the same size as those generated in the style files provided for $\mathcal{A}\mathcal{M}\mathcal{S}$-LaTeX or $\mathcal{A}\mathcal{M}\mathcal{S}$-TeX.

Information on the backlog for this journal can be found on the AMS website starting from `http://www.ams.org/memo`.

A Consent to Publish is required before we can begin processing your paper. After a paper is accepted for publication, the Providence office will send a Consent to Publish and Copyright Agreement to all authors of the paper. By submitting a paper to the *Memoirs*, authors certify that the results have not been submitted to nor are they under consideration for publication by another journal, conference proceedings, or similar publication.

Information for Authors

Memoirs is an author-prepared publication. Once formatted for print and on-line publication, articles will be published as is with the addition of AMS-prepared frontmatter and backmatter. Articles are not copyedited; however, confirmation copy will be sent to the authors.

Initial submission. The AMS uses Centralized Manuscript Processing for initial submissions. Authors should submit a PDF file using the Initial Manuscript Submission form found at `www.ams.org/submission/memo`, or send one copy of the manuscript to the following address: Centralized Manuscript Processing, MEMOIRS OF THE AMS, 201 Charles Street, Providence, RI 02904-2294 USA. If a paper copy is being forwarded to the AMS, indicate that it is for *Memoirs* and include the name of the corresponding author, contact information such as email address or mailing address, and the name of an appropriate Editor to review the paper (see the list of Editors below).

The paper must contain a *descriptive title* and an *abstract* that summarizes the article in language suitable for workers in the general field (algebra, analysis, etc.). The *descriptive title* should be short, but informative; useless or vague phrases such as "some remarks about" or "concerning" should be avoided. The *abstract* should be at least one complete sentence, and at most 300 words. Included with the footnotes to the paper should be the 2010 *Mathematics Subject Classification* representing the primary and secondary subjects of the article. The classifications are accessible from `www.ams.org/msc/`. The Mathematics Subject Classification footnote may be followed by a list of *key words and phrases* describing the subject matter of the article and taken from it. Journal abbreviations used in bibliographies are listed in the latest *Mathematical Reviews* annual index. The series abbreviations are also accessible from `www.ams.org/msnhtml/serials.pdf`. To help in preparing and verifying references, the AMS offers MR Lookup, a Reference Tool for Linking, at `www.ams.org/mrlookup/`.

Electronically prepared manuscripts. The AMS encourages electronically prepared manuscripts, with a strong preference for $\mathcal{A}\mathcal{M}\mathcal{S}$-LaTeX. To this end, the Society has prepared $\mathcal{A}\mathcal{M}\mathcal{S}$-LaTeX author packages for each AMS publication. Author packages include instructions for preparing electronic manuscripts, samples, and a style file that generates the particular design specifications of that publication series. Though $\mathcal{A}\mathcal{M}\mathcal{S}$-LaTeX is the highly preferred format of TeX, author packages are also available in $\mathcal{A}\mathcal{M}\mathcal{S}$-TeX.

Authors may retrieve an author package for *Memoirs of the AMS* from `www.ams.org/journals/memo/memoauthorpac.html`. The *AMS Author Handbook* is available in PDF format from the author package link. The author package can also be obtained free of charge by sending email to tech-support@ams.org or from the Publication Division, American Mathematical Society, 201 Charles St., Providence, RI 02904-2294, USA. When requesting an author package, please specify $\mathcal{A}\mathcal{M}\mathcal{S}$-LaTeX or $\mathcal{A}\mathcal{M}\mathcal{S}$-TeX and the publication in which your paper will appear. Please be sure to include your complete mailing address.

After acceptance. The source files for the final version of the electronic manuscript should be sent to the Providence office immediately after the paper has been accepted for publication. The author should also submit a PDF of the final version of the paper to the editor, who will forward a copy to the Providence office.

Accepted electronically prepared files can be submitted via the web at www.ams.org/submit-book-journal/, sent via FTP, or sent on CD to the Electronic Prepress Department, American Mathematical Society, 201 Charles Street, Providence, RI 02904-2294 USA. TeX source files and graphic files can be transferred over the Internet by FTP to the Internet node ftp.ams.org (130.44.1.100). When sending a manuscript electronically via CD, please be sure to include a message indicating that the paper is for the *Memoirs*.

Electronic graphics. Comprehensive instructions on preparing graphics are available at www.ams.org/authors/journals.html. A few of the major requirements are given here.

Submit files for graphics as EPS (Encapsulated PostScript) files. This includes graphics originated via a graphics application as well as scanned photographs or other computer-generated images. If this is not possible, TIFF files are acceptable as long as they can be opened in Adobe Photoshop or Illustrator.

Authors using graphics packages for the creation of electronic art should also avoid the use of any lines thinner than 0.5 points in width. Many graphics packages allow the user to specify a "hairline" for a very thin line. Hairlines often look acceptable when proofed on a typical laser printer. However, when produced on a high-resolution laser imagesetter, hairlines become nearly invisible and will be lost entirely in the final printing process.

Screens should be set to values between 15% and 85%. Screens which fall outside of this range are too light or too dark to print correctly. Variations of screens within a graphic should be no less than 10%.

Any graphics created in color will be rendered in grayscale for the printed version unless color printing is authorized by the Managing Editor and the Publisher. In general, color graphics will appear in color in the online version.

Inquiries. Any inquiries concerning a paper that has been accepted for publication should be sent to memo-query@ams.org or directly to the Electronic Prepress Department, American Mathematical Society, 201 Charles St., Providence, RI 02904-2294 USA.

Editors

This journal is designed particularly for long research papers, normally at least 80 pages in length, and groups of cognate papers in pure and applied mathematics. Papers intended for publication in the *Memoirs* should be addressed to one of the following editors. The AMS uses Centralized Manuscript Processing for initial submissions to AMS journals. Authors should follow instructions listed on the Initial Submission page found at www.ams.org/memo/memosubmit.html.

Managing Editor: Henri Darmon, Department of Mathematics, McGill University, Montreal, Quebec H3A 0G4, Canada; e-mail: darmon@math.mcgill.ca

1. GEOMETRY, TOPOLOGY & LOGIC

Coordinating Editor: Richard Canary, Department of Mathematics, University of Michigan, Ann Arbor, MI 48109-1043 USA; e-mail: canary@umich.edu

Algebraic topology, Michael Hill, Department of Mathematics, University of California Los Angeles, Los Angeles, CA 90095 USA; e-mail: mikehill@math.ucla.edu

Logic, Mariya Ivanova Soskova, Department of Mathematics, University of Wisconsin–Madison, Madison, WI 53706 USA; e-mail: msoskova@math.wisc.edu

Low-dimensional topology and geometric structures, Richard Canary

2. ALGEBRA AND NUMBER THEORY

Coordinating Editor: Henri Darmon, Department of Mathematics, McGill University, Montreal, Quebec H3A 0G4, Canada; e-mail: darmon@math.mcgill.ca

Algebra, Radha Kessar, Department of Mathematics, City, University of London, London EC1V 0HB, United Kingdom; e-mail: radha.kessar.1@city.ac.uk

Algebraic geometry, Lucia Caporaso, Department of Mathematics and Physics, Roma Tre University, Largo San Leonardo Murialdo, I-00146 Rome, Italy; e-mail: LCedit@mat.uniroma3.it

Analytic number theory, Lillian B. Pierce, Department of Mathematics, Duke University, 120 Science Drive Box 90320, Durham, NC 27708 USA; e-mail: pierce@math.duke.edu

Arithmetic geometry, Ted C. Chinburg, Department of Mathematics, University of Pennsylvania, Philadelphia, PA 19104-6395 USA; e-mail: ted@math.upenn.edu

Commutative algebra, Irena Peeva, Department of Mathematics, Cornell University, Ithaca, NY 14853 USA; e-mail: irena@math.cornell.edu

Number theory, Henri Darmon

3. GEOMETRIC ANALYSIS & PDE

Coordinating Editor: Alexander A. Kiselev, Department of Mathematics, Duke University, 120 Science Drive, Rm 117 Physics Bldg, Durham, NC 27708 USA; e-mail: kiselev@math.duke.edu

Differential geometry and geometric analysis, Ailana M. Fraser, Department of Mathematics, University of British Columbia, 1984 Mathematics Road, Room 121, Vancouver BC V6T 1Z2, Canada; e-mail: afraser@math.ubc.ca

Harmonic analysis and partial differential equations, Monica Visan, Department of Mathematics, University of California Los Angeles, 520 Portola Plaza, Los Angeles, CA 90095 USA; e-mail: visan@math.ucla.edu

Partial differential equations and functional analysis, Alexander A. Kiselev

Real analysis and partial differential equations, Joachim Krieger, Bâtiment de Mathématiques, École Polytechnique Fédérale de Lausanne, Station 8, 1015 Lausanne Vaud, Switzerland; e-mail: joachim.krieger@epfl.ch

4. ERGODIC THEORY, DYNAMICAL SYSTEMS & COMBINATORICS

Coordinating Editor: Vitaly Bergelson, Department of Mathematics, Ohio State University, 231 W. 18th Avenue, Columbus, OH 43210 USA; e-mail: vitaly@math.ohio-state.edu

Algebraic and enumerative combinatorics, Jim Haglund, Department of Mathematics, University of Pennsylvania, Philadelphia, PA 19104 USA; e-mail: jhaglund@math.upenn.edu

Probability theory, Robin Pemantle, Department of Mathematics, University of Pennsylvania, 209 S. 33rd Street, Philadelphia, PA 19104 USA; e-mail: pemantle@math.upenn.edu

Dynamical systems and ergodic theory, Ian Melbourne, Mathematics Institute, University of Warwick, Coventry CV4 7AL, United Kingdom; e-mail: I.Melbourne@warwick.ac.uk

Ergodic theory and combinatorics, Vitaly Bergelson

5. ANALYSIS, LIE THEORY & PROBABILITY

Coordinating Editor: Stefaan Vaes, Department of Mathematics, Katholieke Universiteit Leuven, Celestijnenlaan 200B, B-3001 Leuven, Belgium; e-mail: stefaan.vaes@wis.kuleuven.be

Functional analysis and operator algebras, Stefaan Vaes

Harmonic analysis, PDEs, and geometric measure theory, Svitlana Mayboroda, School of Mathematics, University of Minnesota, 206 Church Street SE, 127 Vincent Hall, Minneapolis, MN 55455 USA; e-mail: svitlana@math.umn.edu

Probability theory and stochastic analysis, Davar Khoshnevisan, Department of Mathematics, The University of Utah, Salt Lake City, UT 84112 USA; e-mail: davar@math.utah.edu

Selected Published Titles in This Series

1316 **Ulrich Bunke and David Gepner,** Differential Function Spectra, the Differential Becker-Gottlieb Transfer, and Applications to Differential Algebraic K-Theory, 2021

1315 **S. Grivaux, É. Matheron, and Q. Menet,** Linear Dynamical Systems on Hilbert Spaces: Typical Properties and Explicit Examples, 2021

1314 **Pierre Albin, Frédéric Rochon, and David Sher,** Resolvent, Heat Kernel, and Torsion under Degeneration to Fibered Cusps, 2021

1313 **Paul Godin,** The 2D Compressible Euler Equations in Bounded Impermeable Domains with Corners, 2021

1312 **Patrick Delorme, Pascale Harinck, and Yiannis Sakellaridis,** Paley-Wiener Theorems for a p-Adic Spherical Variety, 2021

1311 **Lyudmila Korobenko, Cristian Rios, Eric Sawyer, and Ruipeng Shen,** Local Boundedness, Maximum Principles, and Continuity of Solutions to Infinitely Degenerate Elliptic Equations with Rough Coefficients, 2021

1310 **Hiroshi Iritani, Todor Milanov, Yongbin Ruan, and Yefeng Shen,** Gromov-Witten Theory of Quotients of Fermat Calabi-Yau Varieties, 2021

1309 **Jérémie Chalopin, Victor Chepoi, Hiroshi Hirai, and Damian Osajda,** Weakly Modular Graphs and Nonpositive Curvature, 2020

1308 **Christopher L. Douglas, Christopher Schommer-Pries, and Noah Snyder,** Dualizable Tensor Categories, 2020

1307 **Adam R. Thomas,** The Irreducible Subgroups of Exceptional Algebraic Groups, 2020

1306 **Kazuyuki Hatada,** Hecke Operators and Systems of Eigenvalues on Siegel Cusp Forms, 2020

1305 **Bogdan Ion and Siddhartha Sahi,** Double Affine Hecke Algebras and Congruence Groups, 2020

1304 **Matthias Fischmann, Andreas Juhl, and Petr Somberg,** Conformal Symmetry Breaking Differential Operators on Differential Forms, 2020

1303 **Zhi Qi,** Theory of Fundamental Bessel Functions of High Rank, 2020

1302 **Paul M. N. Feehan and Manousos Maridakis,** Łojasiewicz-Simon Gradient Inequalities for Coupled Yang-Mills Energy Functionals, 2020

1301 **Joachim Krieger,** On Stability of Type II Blow Up for the Critical Nonlinear Wave Equation in $\mathbb{R}^{3+1}$, 2020

1300 **Camille Male,** Traffic Distributions and Independence: Permutation Invariant Random Matrices and the Three Notions of Independence, 2020

1299 **David M. J. Calderbank, Michael G. Eastwood, Vladimir S. Matveev, and Katharina Neusser,** C-Projective Geometry, 2020

1298 **Sy David Friedman and David Schrittesser,** Projective Measure Without Projective Baire, 2020

1297 **Jonathan Gantner,** Operator Theory on One-Sided Quaternion Linear Spaces: Intrinsic S-Functional Calculus and Spectral Operators, 2020

1296 **Christophe Cornut,** Filtrations and Buildings, 2020

1295 **Lisa Berger, Chris Hall, René Pannekoek, Jennifer Park, Rachel Pries, Shahed Sharif, Alice Silverberg, and Douglas Ulmer,** Explicit Arithmetic of Jacobians of Generalized Legendre Curves Over Global Function Fields, 2020

1294 **Jacob Bedrossian, Pierre Germain, and Nader Masmoudi,** Dynamics Near the Subcritical Transition of the 3D Couette Flow I: Below Threshold Case, 2020

1293 **Benjamin Jaye, Fedor Nazarov, Maria Carmen Reguera, and Xavier Tolsa,** The Riesz Transform of Codimension Smaller Than One and the Wolff Energy, 2020

For a complete list of titles in this series, visit the
AMS Bookstore at **www.ams.org/bookstore/memoseries/**.